# THE SIMPLE UNIVERSE

BY
IAN BREWSTER & KEN SHIWRAM

# ACKNOWLEDGMENTS

Thanks go out to the following people and organizations for their assistance:

Brian Greene, Professor Lilly Kahn, Roger Penrose, Professor Bob McDonald, Natasha Eloi, Richard Green, Diana Leah, Warren Trudeau, Peter Clancy, Jennifer Baker, Mohammed E. Khan, Walter Brooke, Frank Drake, Katrina Monroe, Professor Paul Curtis, Stacy Whitmore, Cindy Tyson Carter, Dawn Green, Paul Ross, Carla Regal, Katia Holland, Evan Merrin, Dakota Washington, Connie Mills, Lisa Heston, Lisa Howard, Kent Hays and Carl Thompson. NASA, University of Toronto, Creo Media, Massachusetts Institute of Technology (MIT), The Hubble Institute Archive, Official Superstring Web Page, Fermi Laboratories, and SPAR Aerospace.

Personal thanks go to the following individuals. Each of them has made this work possible in their own way:

Nicole Haire, Victoria Brewster, Mark McGovern, Yvonne Bialowas, Katia Corriveau, Matt Thornton, Stephen Boyd, Moe Zazi, Ken James, Christopher Gatsis, Demi Barbito, Joe Michael Linsner, Jennifer Baker, Tanya Baker, Jaime Baker, Mark Baker, Kelvin DeSoto, Debra Austin Lee, Kathleen Noble, Greg Burns, Ellen Trudeau, Miriam C. McDonald, Kevin Frape, Helena Kennedy, Reisa Slade, Rita Shiwram, Justin Galal, Colin Jameson, Sean Madigan, Chad Power, Shannon Haydon, Andy Offield, and Kap Shiwram.

Special thanks to Miss Sara Poirier of the Ontario Science Center for her time, support and advice. You are a very unique and gifted individual.

Very special thanks to Mr. Timothy Ferris for his assistance and answering all my silly questions and to Ken for putting all of this together.

All rights reserved under article two of the Berne Copyright Convention (1971).
We acknowledge the financial support of the Government of Canada through the
Book Publishing Industry Development Program for our publishing activities.
Published by Collector's Guide Publishing Inc., Box 62034,
Burlington, Ontario, Canada, L7R 4K2
Printed and bound in Canada
THE SIMPLE UNIVERSE
by Ian Brewster & Ken Shiwram
ISBN 1-894959-11-6
ISSN 1496-6921
Apogee Books Space Series No. 41
©2003 Apogee Books

# PREFACE

## By Mark McGovern, Ph.D.

The stars have been around a long time. Our ancestors relied on the stars for navigation, to explain the past, to predict the future (astrology), and to look for answers. Our predecessors knew that the stars were created and empowered by something far more powerful and enduring than any fire that could be created on Earth. Many believed that the stars were created by a supernatural being (or beings), and wondered if perhaps that same force had also created them. Creationism says that God created the universe and all of the physical laws that govern events within it. According to that view, God is perfect, eternal, and He existed before the universe began. Our ancestors believed in a static and unchanging universe, where everything is perfectly ordered, because anything less would be less than perfect. Modern science, however, has shown us that the universe is in fact continually changing and evolving. Even the stars are now known to undergo birth, adolescence, old age and death. That the universe (or anything within it) might be permanent and unchanging is contradictory to our current cultural beliefs – if we can't live forever, then neither shall anything else.

The scientific point of view, on the other hand, is (mostly) independent of religious and cultural biases, and attempts to describe reality without reliance on any supernatural beings or events. The methods and thought processes of science are rooted in a branch of philosophy called "reductionism," which means that we come to understand a complex item or event by reducing it to its basic components, which are more easily understood. Science is about things that we can measure, manipulate and test; and "perfection" could then be defined as the condition that exists when science can explain everything perfectly and completely. Prerequisite for such a state of perfection would be a "Theory of Everything" (TOE).

Science, in general, is a very good tool for understanding the physical workings of things. However, as we probe deeper into the microscopic heart of matter, and further back into time, we find that modern science cannot answer such basic questions as "what existed before the universe began?"; "who are we and where did we come from?" and "where are we going?" Science relies on testing and measuring, but we can never build a machine or scientific tool, no matter how large and powerful, that could reproduce conditions as they were at the moment of creation. Although science has conclusively shown us what rocks, plants, animals and people are made of, the entire universe is not like a clock that we can understand simply by dismantling it and examining its parts under a microscope. There are even questions so complex that it is beyond our ability to propose ways to test them. For example, it is theoretically possible for multiple universes to exist, perhaps even an infinite number of them. This is a possibility whose implications are too extensive and complex to hold in the mind, let alone attempt to prove or disprove. It may never be possible to travel to these other universes (if they exist), or even detect them with any sort of scientific equipment. And even if we could travel to another universe, its physical laws might be very different from those we take for granted in our own universe.

A recurring problem in science is the situation where two different, prominent theories are at odds, such as gravity and quantum theory – each is firmly entrenched, but they in some respects contradict one another. Another example is the catalogue of subatomic particles, which may or may not be complete, depending on whose theories you believe. So, what's a poor scientist to do? Some of history's most famous scientists have struggled with the philosophical implications of their work, and were sometimes frustrated that the scientific method could not solve all of their dilemmas. We've come a long way towards understanding "everything," but we're not there yet.

We should, however, keep things in perspective; many useful things have arisen from our imperfect understanding of the universe. We have electronics, lasers, wireless communications, new medicines and myriad other technologies that have indirectly been generated by scientists developing new tools during their quest to find the Theory of Everything. As an example, biologists struggled for a long time to understand the mechanisms of heredity, before they discovered the four chemical bases integral to DNA. Biologists in the early 1940's couldn't propose a theory of how DNA replicated itself until Watson and Crick used X-ray crystallography data to visualize the 3-dimensional double-helical structure of DNA, which immediately suggested a replicating mechanism. How are X-rays generated? – by the same basic technique that's used in the most complex particle accelerators: electrons are slammed into a metal target, and X-rays are the result. How are DNA and the universe related? A major question in the ongoing efforts to attempt to understand the universe is whether life has ever developed on other planets, and if it has, is that life DNA-based like our own? The Theory of Everything becomes even more significant when we include the question of life in the universe.

If you have an interest in learning more about the universe, *The Simple Universe* is a good place to start, because it takes a very complex subject and makes it understandable. Many books dealing with this subject include very few graphic illustrations, but provide pages and pages of words that make a tough subject tougher. Other books of this type will give you the feeling that subatomic physics and astronomy are only about cataloguing the universe, or explaining it all away, and therefore of no interest (unless you're a scientist). The authors of *The Simple Universe* show you that the universe is not just evolving around us, but is also within us. Very philosophical; and perhaps this is the most important part of this book. Many authors discuss physics and religion together with either a creationist or atheist agenda, but fortunately, this book is not spoiled by personal prejudice. The authors discuss why some believe that God is inseparable from the universe, and why others believe that random chance alone may have created the universe, both without preaching to the reader. Instead, the authors seek to enlighten you by providing the all-important historical context of humanity's never-ending quest to understand the "big questions." What may indeed matter most is our human desire to understand the universe, our place in it, and our relationships with one another. The authors of *The Simple Universe* skillfully highlight what the world was like before and after Galileo's observations of the skies, providing an historical and cultural perspective, without attempting to prescribe any answers. The authors provide you with the elegant and valuable insight – that we are part of the universe – by giving you a degree of empathy for Galileo and a new perspective. They have shown you the door that leads to enlightenment, but you are the one who must walk through it.

Some of us have the impression that scientists are people who wear white lab coats and spend their days "in a vacuum," separated from the rest of society, thinking up obscure theories. My experience in science, and in spending time with other scientists, has been very different from that. In 1995, I had the rare opportunity and privilege to speak personally to 11 Nobel Prize winners in Toronto at a very unique meeting (during the inauguration of the John C. Polanyi Chair in Chemistry, at the University of Toronto). Many of these esteemed Nobel Laureates have a deep interest in the society in which they live and do their work. John Polanyi (Nobel Prize in Chemistry, 1986) spends a great deal of his time teaching how science can benefit society. Christian René de Duve, (Nobel Prize in Medicine, 1974) has written philosophically about life and its relationship to the universe in a book titled *Vital Dust: The Cosmic Imperative*. Henry Way Kendall (Nobel Prize in Physics, 1990, for the discovery of quarks) was a founding member of the Union of Concerned Scientists during the Cold War. These and many other scientists are concerned with human society and how we can improve our way of life.

We are mainly visual learners, and *The Simple Universe* uses excellent graphics to generate perspective. The graphics help you to understand the concepts and transform a subject that is very dense and theoretical into something that we can all relate to. The graphic images will endure in your mind, and you will begin to reexamine the world around you, as you begin to see new shapes, structures and patterns. Rather than lose the reader deep in the subatomic forest, the authors have considerately provided a colourful appendix. There's even a glossary for quickly finding definitions of unfamiliar terms. You can read this book in a single afternoon, but I don't recommend that you do. Take your time; read a few chapters each day, and reflect on what you've learned. Share your new knowledge with a friend, and you'll learn more yourself. I hope this experience will be another great step towards your own understanding of the universe and our place in it.

About the author of the preface . . .

Mark McGovern Ph.D. received his doctorate in chemistry from the University of Toronto in 1997. He has written 10 peer-reviewed scientific publications on the subjects of biosensors and interfacing DNA and devices; co-authored a technical book chapter on DNA micro-array scanners; and previously worked in the biotechnology industry. Dr. McGovern is currently employed in the pharmaceutical industry and lives in Toronto, Ontario, Canada.

# DEDICATION

This work is dedicated Michelle Christina Forbes. You were taken from us before your time. I miss you with all my heart. I hope I have made you proud my dearest friend. I love you Michelle, I always have, I always did, I always will!

This work is also dedicated to Amber Forbes. God willing you will grow up in a new world where racism, sexism, homophobia, fear, ignorance and hatred will be replaced by compassion, understanding, tolerance, mercy and justice. If I can leave only one thing behind for you and those children yet to be it is a world filled with hope, a world where you and all that live will be judged by the content of their character.

*For those who seek the truth; no matter where it leads and whom it may offend.*
*For in the end there is only one truth and all knowledge ultimately means self-knowledge!*

# CONTENTS

The universe is the machine, if you like to put it that way, that created you. Now by that I mean the following: every single atom in your body was once inside a star. We are all then brothers in that sense.

— Allan Sandage, Astronomer


To know the atom we must know the universe. To know the universe we must know the atom.

— Timothy Ferris, Physicist


To see the world in a grain of sand.

— Blake, Auguries of Innocence


All that is in tune with thee, O universe, is in tune with me!

— Marcus Aurelius, Meditations


One God, one law, one element
And one far-off divine event,
To which the whole creation moves.

— Lord Alfred Tennyson


What interests me is whether God had any choice in the creation of the universe.

— Albert Einstein, Physicist / Mathematician

God is subtle not malicious, that nature though difficult to understand should be at the root simple and beautiful.

— Albert Einstein, Physicist / Mathematician

We do not really see the creator (God) twiddling 20 knobs to set twenty parameters to the universe as we know it; that's too many. Ever since the Greeks started us on this road to understanding the atom, the fundamental building blocks of the universe, we have had this prejudice that there is a simplicity underneath all of this. There is a deep feeling that the universe must be simple and beautiful.

— Leon Lederman, Physicist

We do not know that it is true, that there is a beautiful underlying theory. We don't even know if, as a species, we are smart enough to know what it really is. We do know that if we do not assume that there is a beautiful underlying theory and do not assume that as a species we are smart enough to find it, we never will.

— Steven Weinberg, Physicist

Finding the TOE (Theory of Everything) will not destroy anyone's faith in God or the belief in a creator; in fact, it shall strengthen our beliefs. It will confirm what the holy scriptures from so many religions have been saying all along. The universe began in simplicity and blossomed into the complex entity we see before us and in us. The universe is testimony to the divine truth of the complexity the creator can bring from simplicity.

— Lilly Khan, Astrophysicist

To my mind there must be at the bottom of it all not some utterly simple equation but an utterly simple idea. To me, that idea, when we finally discover it, will be so compelling, so inevitable and so beautiful, that we shall all say to each other how could it have been otherwise?

— John Archibald Wheeler, Physicist

# CHAPTER 1: MODERN PHYSICS

From the middle of the 20th century until the present day, physics has used as it's foundation two powerful predictive theories in order to understand the universe. These two theories are general relativity, which was conceived by Albert Einstein, and quantum theory (or quantum physics). General relativity is used to describe the workings of the universe on the largest scales, the movement of planets, moons, stars, galaxies, etc. Quantum theory, on the other hand, focuses on the ultra small, the world of atoms and subatomic particles. Both of these theories are powerful predictive tools and have successfully been used to explain much of the universe to us. However, when these two theories are brought together they are found to be in some aspects contradictory, and fail to make accurate predictions of the workings of the universe.

Physicists today, as did Einstein, firmly believe that there is an unseen underlying symmetry to our universe. This symmetry, if properly understood, would help to bind general relativity to quantum theory, and help to develop a solid framework with which we could answer some of the more esoteric questions about the universe. This framework would provide a simple, elegant set of rules capable of predicting the behavior of our complex universe, on the astronomical scale, at the subatomic level, and everywhere in between. Modern physics has undertaken a dedicated search for a theory that will knit general relativity and quantum theory into one unified field theory.

What you are about to read is an overview tracing the search for a unified theory. This search is the story of modern physics, and it has led to some fascinating discoveries about the nature of our universe and our place in it. As you will see, attempts at unification have led to some truly fantastic wonders, which in the past would have been completely beyond conception, let alone experimental observation.

The world we see around us can be described by two fundamental theories of nature – general relativity and quantum theory. These two theories help to explain everything we see, from everyday events like sunsets and rainbows, to astronomical phenomena such as black holes and galaxies.

Although, on the surface, general relativity and quantum theory don't appear to have anything in common, it is believed that if we could raise the average temperature of the universe sufficiently, the two theories could be merged into a common framework. Experimentation has already demonstrated some aspects of this

merging. And cosmologists believe that 15 billion years ago, when the universe was created following the Big Bang, there was indeed enough heat energy in the universe for general relativity and quantum theory to merge. This leads us to believe that finding a unified field theory will not only help us to understand today's universe, but may also give us further insights into its origin.

# CHAPTER 2: DREAMS OF SIMPLICITY

Albert Einstein, regarded by many as one of science's most brilliant thinkers, dreamed of finding what he called a unified field theory – a single equation that would account for and explain every fundamental process found in nature, from the behavior of matter and energy inside the atom to the motions of the wheeling galaxies hurling through space. Today scientists are very close to grasping Einstein's dream of a unified field theory. Physicists studying the properties and interactions of atoms have accumulated a growing body of evidence pointing to an underlying simplicity that governs the entire universe. Both theory and experimentation indicate that the universe began in a state of perfect simplicity, and evidence of this simplicity can be found in the heart of every atom.

The search for a unified field theory has led cosmologists to consider the farthest reaches of space. Astronomers have created advanced methods and machines to look deeper into space than ever before. The most notable addition in recent years is, without a doubt, the Hubble Space Telescope. The Hubble has allowed us to see things not encountered by any other telescope or probe. What the layman might not realize, however, is that the further into space we look, the further back into time we are looking. The light from a star 10 billion light-years away took 10 billion years to reach us, so what we are seeing is that star as it was 10 billion years ago. This ability for advanced telescopes to see deep into space and time is helping us to better understand the origin of the universe.

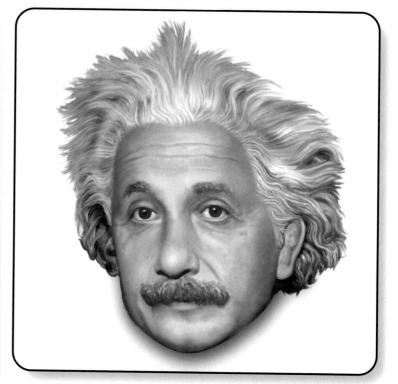

We know that light is, hands down, the fastest thing in the visible universe. Light traveling through the vacuum of space moves at an astonishing 186,000 miles per second. The speed of light through air, water, glass, or any other transparent medium is somewhat less than that, but not appreciably so. Even at that fantastic speed, light still takes time to cross astronomical distances. The distances to even the nearest stars are large, cumbersome numbers when expressed in miles, so astronomical distances are declared in light-years (the distance that light will travel in one year at 186,000 miles/second, which is 5,865,696,000,000 miles or 9,460,800,000,000 kilometers).

The Moon is a quarter of a million miles from Earth, and light takes approximately 1.3 seconds to reach the Moon from Earth. Although this doesn't sound like much, in normal conversion it is quite noticeable. And, of course, signals to Earth originating on the Moon would also take 1.3 seconds to travel that distance. Radio waves to the Moon travel at the speed of light. So, assume I'm on Earth and you're on the Moon; if I asked you a question and you answered it the moment you heard it, I'd still have to wait 2.6 seconds to get an answer. 2.6 seconds might be tolerable for having a conversation, but it would be an unacceptable delay if you were trying to control machinery or land a rocket.

Light traveling from the sun takes about 8.3 minutes to reach Earth. We on Earth don't actually see the sun as it is, we're always seeing it as it was 8½ minutes earlier. If the sun were to (hypothetically) blow itself to bits this very second, you and I would have 8.3 minutes of comfortable living before the Earth was vaporized. Distances in this range are often expressed in light-minutes, instead of light-years. The distance of Earth from the sun is about 93 million miles, called one "astronomical unit" (or AU), and is equal to 8.3 light-minutes.

The star Proxima Centauri (of the Alpha Centauri triple star system) is the closest (extra-solar) star to Earth; it is 4.22 light-years away. This means that light takes 4.22 years to get from Proxima Centauri to Earth, and that we are seeing it as it was more than four years ago – we are looking four years back in time.

One aspect of studying the universe to pursue our unified field equations is using instruments like the Hubble Telescope to look out into space. But at the same time, the universe is all around us, not just out among the stars and galaxies; we live in the universe. The invisible atoms that make up the matter all around us, and make up us too, have, hidden in them, clues to the origin of the universe. A stone or a drop of water is a galaxy of atoms; and each of those atoms was once inside a star; the particles that make up those atoms can trace their lineage all the way back to the beginning of time. The nuclear fusion in stars is a process that we now understand. From this fusion and the subsequent events in the lives of the stars come more than 100 different atomic elements – the universe was born with hydrogen and helium; the rest of the atoms it has had to make for itself. We can see from this that the stars and the subatomic realm have been allied all along. In pursuing our unified field equations we're trying to understand that alliance completely.

The Andromeda galaxy is a huge star group more than 2,300,000 light-years away from Earth. There are over 300 billion stars that call the Andromeda galaxy home. The light we see from these stars took more than 2.3 million years to reach us. That light started its journey through space before human beings had even evolved and Mastodons still walked the Earth. There are objects even deeper in space, objects whose light takes billions of years to reach us. These objects were created during the early universe and their study may yield clues to its evolution.

## A VIEW OF THE INNER UNIVERSE

Scientists who study the Earth have concluded that our home planet is over 4.5 billion years old. That is a staggering age, but older still are the atoms that make up the Earth, and the air, the plants, animals, and you and me. By looking into the cores of these atoms we can find clues that help us understand the simplicity that underlines the entire universe. The more closely we observe the very small (the atomic and subatomic scales), the more evidence we find of the simplicity of the universe, and the closer we come to completely visualizing its origin.

Life on Earth comes in countless shapes and sizes. The genetic blueprint that this tree followed in its growth is contained in the nucleus of every one of the tree's cells. The functioning of the grown parts of the tree is controlled by that same blueprint.

The tree is covered with hundreds of leaves, which gather sunlight, the energy source that keeps the tree alive.

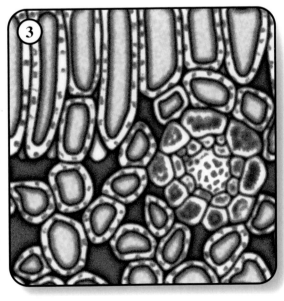

Each leaf is made up of uncountable billions of microscopic cells. Each cell has within it all of the necessary information to copy itself and to function. This genetic information was contained in the seed from which the tree grew, and the seed in turn received it from the parent tree. This genetic heredity function is the same for trees, people, and all other life on Earth.

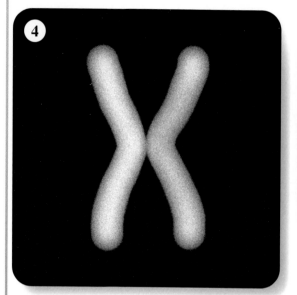

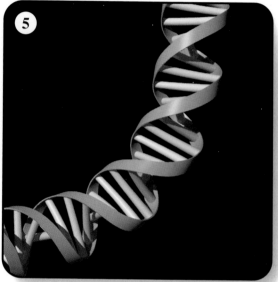

Within the nucleus of each of the leaf's cells are chromosomes. It is within these chromosomes that the genetic information is encoded. Each chromosome is made up of DNA molecules, which are extremely long and so must be coiled and compacted and held in place. The chromosome is the resulting package.

DNA molecules are constructed from two DNA strands linked together like a ladder to form the familiar double helix. Each "rung" of the ladder is a pair of complex carbon-based molecules. There are only four types of molecule used in these rungs, and only two pairs (types of rung) that can exist, but a string of rungs in a specific order makes a "letter" of the genetic alphabet. The combination of letters within a DNA molecule spells out the genetic blueprint. The DNA mechanism is the same in all living things on the Earth; all that changes from one organism to the next is the specific letters in its blueprint.

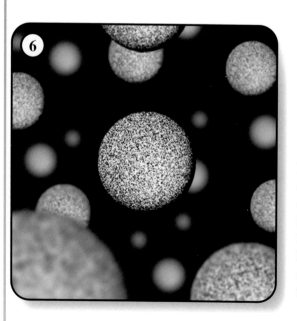

DNA molecules include carbon atoms that are older than the Earth and were formed millions, or even billions of years ago. All atoms of elements heavier than iron (which is the key atom of the hemoglobin molecule in your blood) can only have come into existence when a star exploded in a nova – massive stars had to have been born and then died before there could be life on Earth.

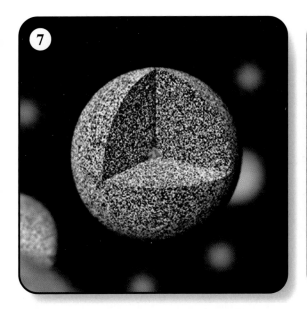

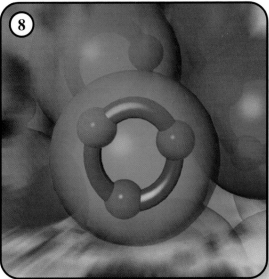

Most of the volume (not mass) of any atom is the space in which its electrons orbit. Any given electron orbit exists at one of multiple energy levels (the details of what an electron "orbit" comprise are very detailed and not included here). The energy difference between two energy levels is exactly equal to the energy of one photon. As we'll see (in Chapter 4) the photon is the carrier of the electromagnetic force. When a photon strikes an atom, it is absorbed and its energy goes to increase an electron to its next energy level. Similarly, when an atom emits a photon, an electron will drop down one energy level.

Deep inside the atom is its nucleus. The nucleus occupies almost none of the atom's volume (the diameter of the atom is 100,000 times the diameter of its nucleus), but it makes up almost all of the atom's mass (the protons and neutrons that make up the nucleus each weigh more than 1800 times what an electron does).

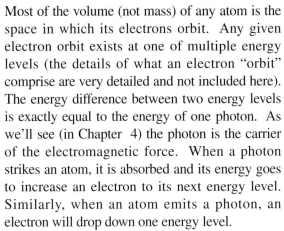

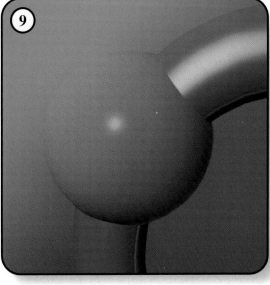

As small as protons and neutrons are, they too are made of even smaller and more fundamental particles, the quarks. At this level we've reached the territory of the nuclear forces, which govern the structure, life and death of the atom, and therefore all matter in the universe.

# CHAPTER 3: THE TOOLS OF THE SEARCH

A great deal of what we know about the structure of the atom comes from particle accelerators. These machines speed up particles to a very high velocity and collide them together. Physicists study the debris that comes flying out of these collisions and consider its make-up and behavior in the context of their theoretical expectations. From this we have slowly over time learned more about the atom and its constituent particles. It may sound like a somewhat crude method of research, but it has worked exceptionally well, replicating high energy experimental environments, resembling the early universe, that we can't create any other way.

Here is a diagram of how the first accelerators operated. The basic principles that govern how this machine operates are still in use today, although modern accelerators are much larger than their predecessors. In simplified form, the diagram shows the operation of a particle accelerator.

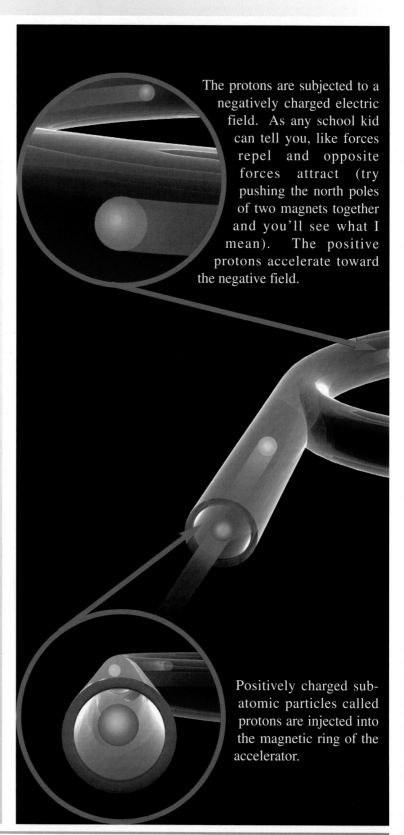

The protons are subjected to a negatively charged electric field. As any school kid can tell you, like forces repel and opposite forces attract (try pushing the north poles of two magnets together and you'll see what I mean). The positive protons accelerate toward the negative field.

Positively charged subatomic particles called protons are injected into the magnetic ring of the accelerator.

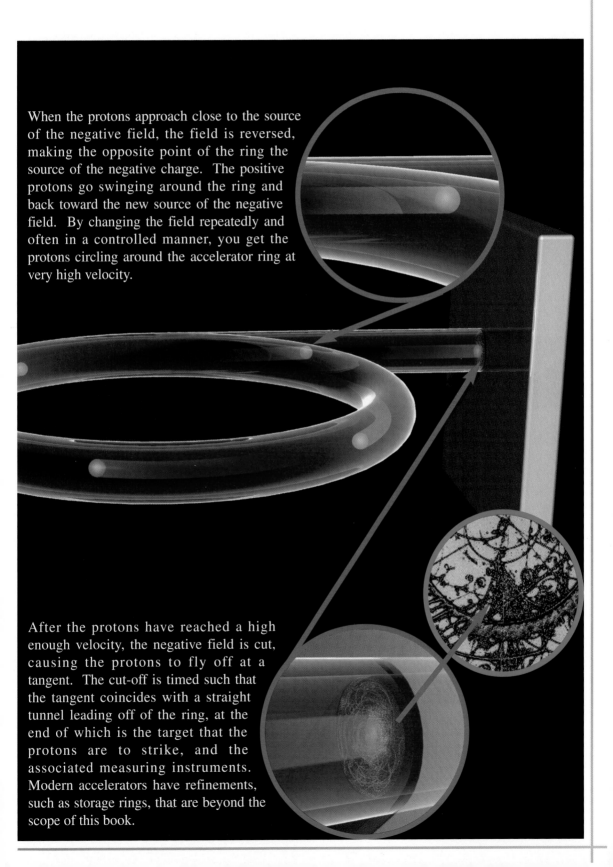

When the protons approach close to the source of the negative field, the field is reversed, making the opposite point of the ring the source of the negative charge. The positive protons go swinging around the ring and back toward the new source of the negative field. By changing the field repeatedly and often in a controlled manner, you get the protons circling around the accelerator ring at very high velocity.

After the protons have reached a high enough velocity, the negative field is cut, causing the protons to fly off at a tangent. The cut-off is timed such that the tangent coincides with a straight tunnel leading off of the ring, at the end of which is the target that the protons are to strike, and the associated measuring instruments. Modern accelerators have refinements, such as storage rings, that are beyond the scope of this book.

## THE FERMI PARTICLE ACCELERATION LAB

The first particle accelerators were built in the United States during the 1930s. These first accelerators were rather small; they could fit inside your hand. But particle accelerators have been getting bigger since the '30s and today the largest ones in the world have rings that are miles long. The world's largest particle accelerator, Fermi Lab, has an accelerator ring that is nearly three miles in circumference.

This is a map of Fermi Lab, out on the Illinois plains in the United States. Fermi Lab is the largest particle accelerator in the world. It is home to countless scientists who are engaged in pure research; their ultimate goal is find the unified field theory envisioned by Einstein so many decades earlier.

### THE ACCELERATOR RING

This is the main ring into which particles are injected at Fermi Lab. This ring is nearly three miles in circumference.

The main research lab and office area at Fermi Lab.

# PARTICLES CRASHING

This is a computer-generated image of the result of a particle collision. Each line represents the path taken by an item of debris (tiny particles that come spinning off from the collision). We cannot actually see these particles – they are far too small – but special detectors can identify their presence.

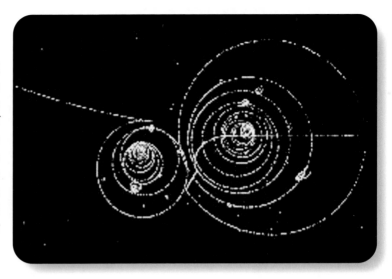

The energy that runs Fermi Lab is about equal to the energy needed to run a small city. Buried deep below Fermi Lab is an accelerator ring whose tunnel is as small as a garden hose. Through this ring particles are accelerated to the speed of light and collided with each other and fixed targets. Miniature explosions created by subatomic particles colliding together, in laboratories such as Fermi Lab, are revealing a wide variety of previously unknown exotic subatomic particles that interact by means of four forces: gravitation, electromagnetism, the nuclear strong force and the nuclear weak force. These particles and forces are the building blocks of everything we see around us.

# CHAPTER 4: THE FOUR FORCES

In studying matter on the very small scale, physicists realize that the make-up of our universe is governed by a small but growing number of subatomic particles and four fundamental forces. The interactions of these forces – gravity, electromagnetism, the nuclear strong force and the nuclear weak force – control everything we see in the universe, from the exotic to the everyday. Most of us are unaware of these forces and how they act upon the universe, let alone how they affect our daily lives. An understanding of what these forces do is essential to formulating a unified field theory (or Theory of Everything) that will account for and describe everything in the universe.

It is difficult to explain even the most basic concepts in science to the average person without a common frame of reference – something that the everyday person can grasp. With the "four force" theory, unlike many advanced theories in physics, we have numerous examples from the everyday world to help explain things in a way that is easily envisioned. If we take the example of playing pool, we can see the four forces in action.

The objects we see around us in our everyday lives, like a pool ball, the pool cue, a rock, a chair, even ourselves, may appear solid, but in reality their solidity is an illusion. At the level of our senses, a cue ball certainly looks solid, but the atoms that make it up are mostly empty space. The apparent solid nature of the cue ball is caused by the electromagnetic force, which binds all of its atoms into a single rigid structure. The atoms of the cue ball are electromagnetically held together much like individual bricks are held together (by mortar) into a single building.

When the cue ball strikes another ball, what we see as two solid balls colliding appears very different on the atomic scale. When seen at the atomic level, the collision of the balls involves an exchange of photons between the atoms of the cue ball and the atoms of the ball being struck. This exchange of photons (an exchange of energy) is the "flow" of the electromagnetic force. The result of the electromagnetic interaction is that the balls are repelled from each other. We give credit to the pool player, but the fundamental force that allows this to happen is electromagnetism. Electromagnetism is infinite in range, and its strength decays proportional to the square of the distance between the interacting objects. Electromagnetism is the force responsible for, among many other things, bringing heat from the sun, light from distant stars, and putting the 8 ball in the side pocket.

Gravity is the universal attraction of all massive particles to one another (where massive means having a non-zero mass; not all subatomic particles have mass). Gravity is the weakest of the four forces and its effects are proportional to the masses involved. Since planets have a lot of mass, we tend to think of gravity as a very strong force. Gravity's effect is infinite in range, and as with electromagnetism, gravity's strength decays proportional to the square of the distance between the interacting objects. Gravity always attracts (never repels). Gravitation is the force that makes all thrown or hit objects (like the pool ball) fall back to Earth (or the pool table in our example). Gravity keeps us pinned to Earth and keeps the stars, planets and galaxies locked in their orbits. Gravity is theoretically carried by particles called gravitons. Although gravitons have never actually been detected, experimental evidence has allowed several of their properties to be determined.

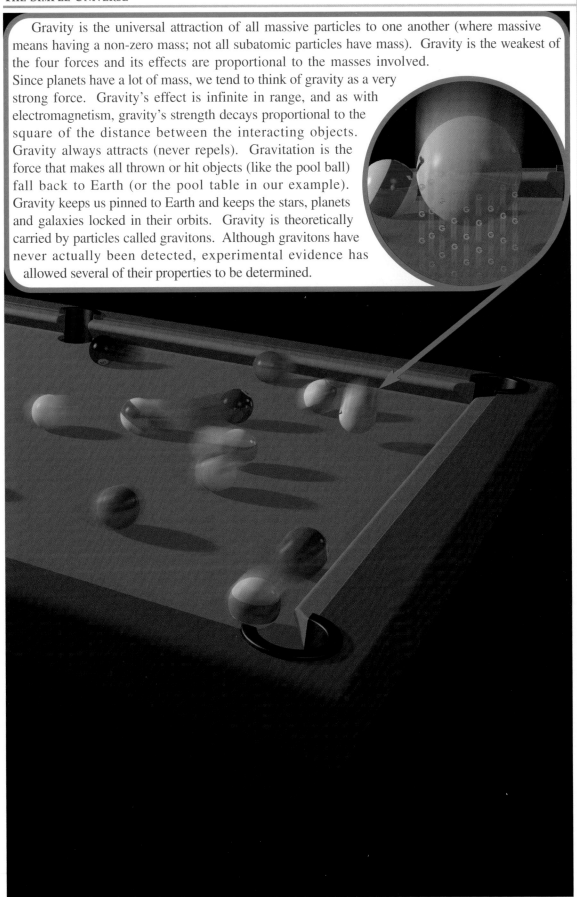

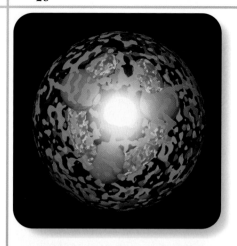

The nuclear weak force is involved in, among other things, the burning of our sun (without it, deuterium fusion could not take place). The nuclear weak force regulates the phenomenon of nuclear decay (whereby a subatomic particle breaks down to a different type of subatomic particle plus emitted energy). The weak force involves exchange particles (force carriers, like the photon) called intermediate vector bosons, or weak bosons. These exchange particles come in three flavors – $W^+$, $W^-$ and $Z^0$ – which have been detected in particle accelerators. The range of the nuclear weak force is very short, limited to the nucleus of the atom and is, in fact, only about 0.1% of the diameter of a proton.

The nucleus of every atom is made up of protons (which are positively charged) and neutrons (which have no charge). The nuclear strong force holds the atom's nucleus together against the enormous forces of repulsion between its protons, which is strong indeed (protons all have a positive electromagnetic charge, and like charges repel). The strong nuclear force has a very short range, about equal to the diameter of the atom's nucleus. In the standard model of the nuclear strong force, subatomic particles called quarks are bound together into protons and neutrons, and the exchange particle is called a gluon. Without the nuclear strong force there could be no protons and neutrons, therefore no atoms, and no stars, planets, rocks, air, or you and me.

Note: The "nuclear weak force" is also referred to as the "weak nuclear force." The same is true for the nuclear strong force. Physicists generally refer to them simply as the weak force and the strong force. For the sake of simplicity, we shall refer to them hereafter as the weak force and the strong force.

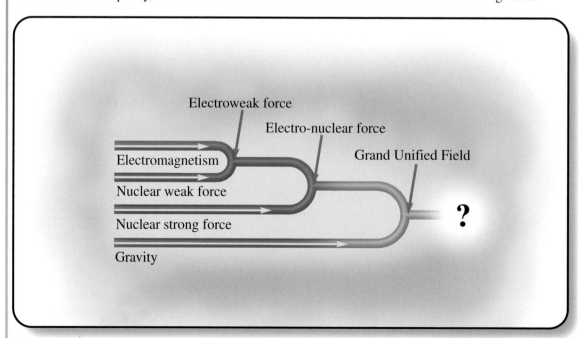

| | FORCE | EXCHANGE PARTICLE | RANGE |
|---|---|---|---|
| ELECTRO-MAGNETISM | 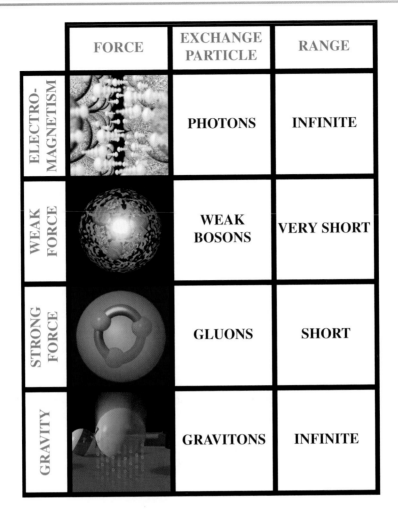 | PHOTONS | INFINITE |
| WEAK FORCE | | WEAK BOSONS | VERY SHORT |
| STRONG FORCE | | GLUONS | SHORT |
| GRAVITY | | GRAVITONS | INFINITE |

## WHY ONLY FOUR FORCES?

Albert Einstein tried in vain until the day he died to find out why there are only four forces and not more, and why those four forces differed so much from one another. Einstein, like other physicists of his day, pondered the apparent complexity of the universe in the hope that a single theory or principle would come forth to explain its workings. Einstein believed that this principle should be, at its root, simple, elegant and beautiful.

# CHAPTER 5: THE SEARCH FOR SIMPLICITY

Einstein was confounded by his search for an ultimate simple truth to the universe because he had a hard time understanding why the four forces had such profound differences. Einstein's mind was flooded with questions that he couldn't answer, such as: "Why are there such things as a positive and negative electrical charge, but no such thing (as far we know) as negative gravity?"; "Why do the strong and weak forces operate only within the confines of the atom, while electro-magnetism and gravity are infinite in range?" These questions, among many others, are what stymied Einstein in his search for a Unified Field Equation. His dilemma was by no means unique, however. The search for simplicity began long before Einstein. The search for a simplified theory that explains the universe is not new, but is an idea which can trace its origin through time all the way back to ancient Greece.

Plato and his pupil Aristotle first proposed atomic theories in ancient Greece. Aristotle proposed that the universe was governed by two forces: gravity, the force that makes objects fall; and levity, the force that makes objects rise.

Lucretius popularized atomic theory after the deaths of Plato and Aristotle. In one of his most famous lectures, Lucretius proposed that the reason why a wedding ring grows thinner over the years, or metal gradually gets weaker over time, was because of tiny indivisible pieces of matter called atoms, which gradually fell off.

During the dark ages, atomic theory all but vanished as the Roman Catholic Church essentially banned any scientific research and experimentation which went against creationist teachings.

Galileo Galilei created what were some of the most powerful telescopes of his day. Galileo found that other planets have moons, that the Milky Way was a vast pasture of stars, and that the Earth and the other planets revolved around the Sun. Galileo helped to marry the Earth to the universe and set the stage for all of the unification theories to come.

Robert Boyle defines specific chemical elements in the mid-17th century.

In 1687 atomic theory resurfaced when Isaac Newton presented his theory of gravity to the world. Newton perceived the universe as being made up of atoms which were powered by the force of gravity.

John Dalton, in 1802, introduces modern atomic theory into chemistry, and forever ties physics to chemistry.

Ernest Rutherford discovers positively charged particles (protons) in center of the atom.

James Chadwick, not long after Rutherford found protons, discovers that the center of the atom is also home to uncharged particles (Neutrons).

In 1873, James Clerk Maxwell proposes for the first time that magnetism is really just an aspect of electricity, thereby taking the first true step toward a unified field theory. The new force was called electromagnetism.

Einstein creates his theory of relativity and greatly enhances our view of what gravity is and how it behaves. Einstein, along with Max Planck, showed that energy comes in tiny indivisible packets called quanta; one quantum is the minimum size that an energy packet can be. He also found that electromagnetic radiation produces a kind of quanta, which he called photons.

It appeared that there were only two forces controlling the universe until Hideki Yukawa (a Japanese physicist) and Enrico Fermi (the mathematician / physicist for whom Fermi Labs is named) identified that there were two more forces, acting within the nucleus of the atom. These new forces were called the nuclear strong force and the nuclear weak force. With their discovery there were now four forces that governed the universe.

Paul Dirac discovers the very bizarre entity antimatter – composed of particles whose properties exactly the same as their normal matter counterparts, except for charge, which is opposite. For example, electrons in normal matter are negatively charged particles, whereas an antimatter electron (a positron) has a positive charge. Dirac found that if a particle of normal matter met up with its twin antimatter particle, the two would completely annihilate one another (releasing an amount of energy determined by Einstein's famous $E=mc^2$). This would prove to have important ramifications in particle physics and the search for a unified field theory.

By the middle of the 20th century, scores of new subatomic particles were being discovered, seemingly almost on a daily basis. Particles such as pions, muons and neutrinos became buzzwords. It appeared that the ongoing search for simplicity was showing exactly the opposite – that the universe was, in fact, much more complex than we had imagined.

However, hope for a unified field theory and a simple understanding of the universe were brought back to the forefront of science when Murray Gell-Mann found that the protons and neutrons within an atom were made up of even smaller particles, which he called quarks. This set the stage for the modern era of unified field theories.

Then Steven Weinberg, along with a team of prominent physicists, proposed that electromagnetism and the weak force may be aspects of a single "electroweak" force. However, they lacked the ability to confirm this theory, at least at the time.

By the beginning of the 1980s, it appeared that physicists were finally within reach of Einstein's coveted goal of a unified field theory. With the creation of string theory and supersymmetry many physicists feel that a unified field theory – a Theory of Everything – might be right around the corner.

# CHAPTER 6: GRAND UNIFIED THEORY: PART ONE

## CONFIRMATION OF THE ELECTROWEAK THEORY

Steven Weinberg, with a team of other physicists, was the first to determine mathematically that electromagnetism and the weak force could be combined into a single electroweak force. They surmised, correctly, that the universe would function more simply if the temperature was turned up. If the temperature of the universe was hot enough, the electromagnetic force and the weak force should come together into a single force, with a single set of properties. Experimentally, there was a problem, however – how do you turn up the temperature enough to confirm that these two forces come together. The answer came from an unusual source – that odd quantity called antimatter, discovered by Paul Dirac.

## SIMPLICITY UNDER HIGH ENERGY!

Electroweak theory considers electromagnetism and the weak force interaction to be different aspects of the same force, and the exchange particles (force carriers) of both come into play in electroweak reactions. Electroweak theory must also account for the complicating factor that the weak force bosons are massive, while photons (the electromagnetic exchange particles) are without mass. The details are rather highly technical, and so are not pursued here. Required is an additional interaction with an otherwise unseen field, called the Higgs field, that pervades all space (See Chapter 19).

If we compare the electromagnetic force to the weak force in our current universe, they are very different. The electromagnetic force is carried by highly energetic massless photons that act over an infinite range. The weak force, on the other hand, is carried by very massive weak bosons, which act only over an extremely short distance (the weak force is limited in range to the nucleus of the atom). It's worth noting, however, that within the small range over which the weak force operates, the strength of the electromagnetic force is comparable to that of the weak force.

Knowledge gained in pursuing a unified field theory suggests that the universe would function much more simply in a high ambient thermal energy state (a high average temperature). Today's theories state that if we could "turn up the heat," so that our forces existed in a higher-energy environment, things would change. When considering the electroweak force, it is also important to realize that at the time when the universe was at a sufficiently high temperature to have exhibited the electroweak force ($1/10$ billionth of a second after the Big Bang), it was also an extremely tiny universe, greatly limiting the amount by which electromagnetism's strength could decay over distance.

At the time when the electroweak force was first proposed, there had as yet been no experimental confirmation of the existence of the exchange particles for the weak force (weak bosons). Three types of weak bosons had been predicted – $W^+$, $W^-$ and $Z^0$. For the sake of simplicity, however, we shall refer to them collectively as the "weak bosons." As we'll see, the weak bosons were not actually detected until experiments reproducing "Electroweak Era" high energy conditions were performed.

In the Electroweak Era of the universe, the ambient thermal energy was high enough, and (more importantly) the total size of the universe was small enough, that weak bosons and photons exhibited similar properties over the limited distances involved, resulting in electromagnetism and the weak force interacting with hadrons and leptons (the mass particles) in the same way (i.e., with a single, shared set of characteristics).

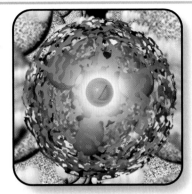

## THE WEAK BOSON PROBLEM

Although the theorized weak bosons (W and Z particles) could describe the weak force and electroweak force interactions on paper, there was one small problem – how do you create an environment with enough energy to actually detect them in action and verify their existence? The weak bosons in our current universe have an undetectably short life span and exist only in combination with other particle types. At temperatures of around $10^{15}$ °K the weak bosons can "stand alone," distinct from other particles and perhaps be detected by their interactions with other particles. However, even the cores of the hottest stars are not that hot, and certainly no lab on Earth could produce that kind of test environment. Physicists were at a standstill, and it appeared that the unified field theory was once again out of reach – or was it?

In 1983, scientists at CERN, a giant particle accelerator on the border between Switzerland and France, came up with a way to produce the energy levels needed to detect the weak bosons.

## THE ANTIMATTER SOLUTION!

Most particle accelerators speed up particles, usually protons, to the speed of light and then smash them into stationary targets. Protons are easy to find – there's at least one proton in the nucleus of every atom. Protons hitting a stationary target (e.g., a lead wall) produce very large subatomic explosions. However, what if you were able to run protons into each other, or, better yet, what if you were able to run protons into their antimatter counterparts, antiprotons?

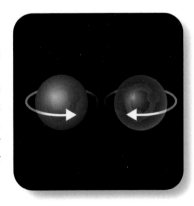

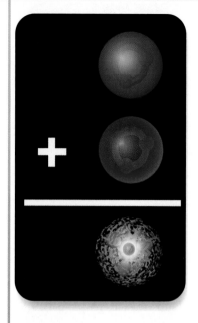

## CREATING THE SPARK

In 1983, Carlo Rubbia, an Italian physicist, conceived a plan to create the ambient energy needed to detect the weak bosons (weak force exchange particles) in action. Carlo believed that colliding matter and antimatter together would produce enough energy (through their mutual annihilation), for a very brief instant, to simulate the conditions of the electroweak era. (Antimatter particles, you'll recall, have the same properties as their matter counterparts except for an opposite electrical charge.) Antimatter is extremely rare in our universe, and is very difficult to create. But using the antimatter generator at CERN, Rubbia and his team smashed antimatter protons into normal matter protons. The resulting explosion, although confined to an invisibly small volume, created more energy in a fraction of a second than the entire output of all the world's generators would in a day. In the recorded behavior of the debris from the matter / antimatter explosion, the weak bosons in action were finally detected.

This is a picture from CERN's detector screen revealing the presence of a weak boson (the tiny red dot at the center of the picture). This was the first real empirical proof that Einstein was correct – that there is an underlying simplicity to the universe which can be explained in simple, rational and elegant terms. Carlo Rubbia shared the 1984 Nobel Prize in Physics for employing antimatter in the experiment to confirm the existence and behavior of the weak bosons. Rubbia brought the scientific world one step closer to Einstein's dream of a unified field equation – a Theory of Everything.

# CHAPTER 7: GRAND UNIFIED THEORY: PART TWO

## ENTER THE X PARTICLE

With the confirmation of the weak bosons and the electroweak force at CERN in 1983, Carlo Rubbia and his team showed that Steven Weinberg was absolutely correct about the relationship between electromagnetism and the weak force. This physical confirmation helped to further stimulate the search for a unified field equation which would explain the entire universe. With electro-magnetism and the weak force combined, physicists reasoned that at even higher energy levels the strong force would combine with the electroweak force to form a new unified force which they called the electro-nuclear force. A new exotic particle was predicted, an exceedingly massive particle which they called the X particle. The X particle (also called the X boson) would be the exchange particle (force carrier) for the unified electro-nuclear force. At the detailed level, theory predicts that an X particle would be able to change a quark into a lepton or an antiquark. The Grand Unified Era is the name given to the time when strong, weak and electromagnetic forces were unified in the universe, a time preceding the inflation period when the universe began to expand.

## ALONG CAME MR. X

With the confirmation of the electroweak force, physicists predicted that, at even higher temperatures, the strong force would unite with the electroweak force to produce a single "electro-nuclear force." It was calculated that the theoretical X particle, the proposed exchange particle for the electro-nuclear force, would be extremely massive by subatomic standards. There is no absolute certainty that an actual new particle is required for this unification – it may be that the properties of the gluon (the strong force exchange particle) and the photon and weak bosons become similar in the extremely hot, almost infinitesimally small universe that existed during the Grand Unified Era when the electro-nuclear force held sway. It may be that both a new particle and the functional meshing of the various exchange particles are required. For the sake of convenience, physicists refer simply to the X particle, and we shall do the same. The ability to generate an experimental environment equivalent to the Grand Unified Era is far beyond the abilities of today's technology.

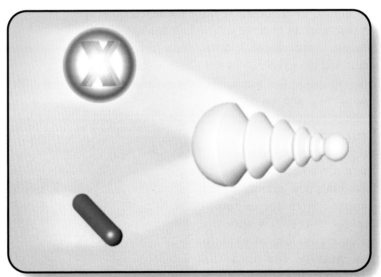

The electro-nuclear force requires a force exchange mechanism – just like the fundamental forces – and physicists refer to it as the X particle. X particles will combine the functions of gluons and photons (above) as well as gluons and weak bosons such that the electromagnetic, weak and strong forces will all three behave in a similar manner, as a single, unified force.

Science has yet to detect an X particle. Scientists in Europe, North America and Japan are feverishly trying to find an X. To date, there have been many false finds, and much evidence implying the presence of this ultra-elusive particle, but no lab has yet been able to positively detect one. Like the weak bosons (W and Z particles), X's are very difficult to detect. This is partially due to the fact that they have extremely short life spans and exist for only tiny fractions of a second before decaying back into energy and more mundane particles. The detection of the weak bosons, along with confidence in the existence of X particles, has resulted in many interesting theories about the universe, some or all of which might be verified by unified field equations. One of these theories suggest that matter, all matter, will one day decay into energy, and so the structure of the universe as we know it will gradually break down. If the universe continued to expand, then energy too will decay to a density indistinguishable from zero – the universe will be truly empty.

As well as being very massive, the theoretical X particles are energy pigs. They require a lot of ambient energy to form. Scientists who measure the energy levels of subatomic particles use a unit of energy called GeVs (Giga Electron Volts, or $10^9$ Electron Volts). Our X particle friend would be in the 115 GeV range; this is an extremely high energy level for a particle!

## AND WHAT ABOUT GRAVITY?

The unified field equations have predicted the combination of three of the four forces into one "superforce." Physicists have confirmed the existence of the electroweak force, dropping the number of forces from four to three, and are on the trail of the hypothesized X particle, carrier of the electro-nuclear force. But what about gravity? There are no tenable theories describing how gravity might combine with the electro-nuclear force. Without a suitable equation to explain how gravity fits into the symmetry of the universe, there can be no true unified field equation. And the nature of gravity is such that we will probably have to experiment away from the surface of the Earth to obtain useful results.

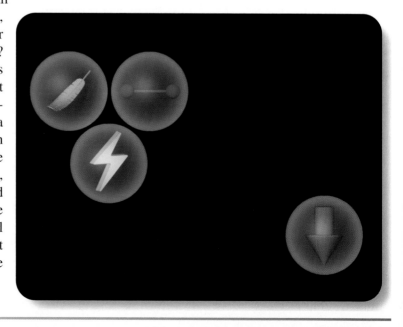

# CHAPTER 8: THE ULTIMATE FATE OF ALL MATTER

The discovery of the electroweak force, and the mathematical proposal of the X particle, have led to some rather startling insights into the nature of our universe. One of the predications made by current unified field theories is that matter may not be a permanent fixture in our universe. According to physics as we understand it today, protons, which with neutrons form the nuclei (cores) of all atoms, will eventually decay. If this is true, then all matter may be nothing more than one phase in the ongoing evolution of energy. Like solid ice forming on a pond in the winter, and then thawing back to water with the spring, perhaps matter is nothing more than a temporary state of energy.

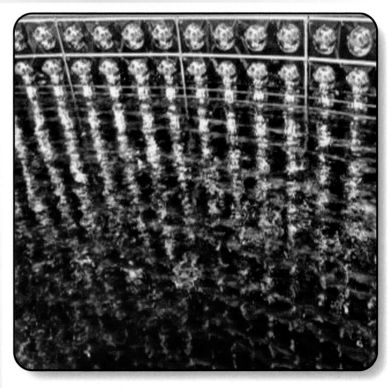

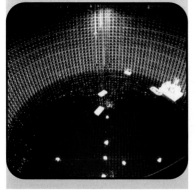

In large, specially constructed water tanks around the world, there are experiments under way to try and prove with empirical evidence that protons do decay. The theory goes something like this: since scientists believe that protons will eventually decay, then if we wait long enough, we should see a proton decay and then die in a flash of radiation. Now, on average, a proton may take billions of years to decay, but if you assemble enough of them in one place, then you just might get lucky.

These pictures, and the one on page 38, are of a specially constructed tank, lined with detectors, looking for the telltale flash of radiation which will signal the decay of a proton. The tank is filled with more than 660 tons of pure water and is hidden a few miles under ground to ensure that cosmic radiation and other particles from space do not interfere with the experiment. Experiments have yet to detect a proton decay. However, most physicists feel fairly confident that the detector tanks in Canada, the US, Europe or Japan will soon yield up a decaying proton. If current unified field theories are correct, then matter, like all other things in our universe, is living on borrowed time and cannot exist forever.

### PROTON DECAY DETECTOR

A specially constructed tank, lined with detectors, looking for the telltale flash of radiation which will signal the decay of a proton.

The apparent fact that all matter is destined to decay is a relatively new theory, brought about by the unified field equations. However, the seeds of this idea were planted 400 years ago by the great astronomer and scientist Galileo Galilei. Galileo lived near Venice, Italy and was a well-respected teacher and scientist. In 1609, he led a small group of local senators up to the top of a tower to have their first look through his new telescope. Telescopes were not a new invention at the time, but Galileo's telescopes were much stronger and more accurate than most. Needless to say, Galileo's telescope was a success and the senators were very impressed. He was given a rather large grant, his tenure at the university where he taught was extended, and they commissioned Galileo to build several more telescopes to be used for spotting ships at sea.

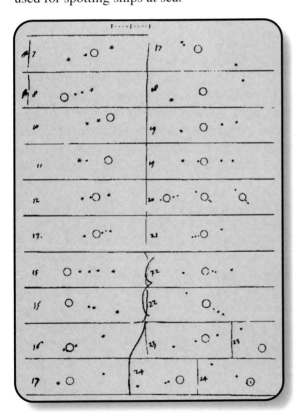

One of Galileo's many hand-drawn diagrams, this is a record of Jupiter and it's orbiting moons taken over several days. Note how the positions of the moons change over a period of only days.

## THE MARRIAGE OF EARTH TO THE UNIVERSE

While others in Venice trained their telescopes on the sea, Galileo used his telescopes to observe the sky; what he saw changed our view of the universe forever. Galileo saw that the Moon was not simply a perfect sphere, but a world rich with craters, mountains and valleys. He also saw that other planets had moons as well, and discovered four moons orbiting the planet Jupiter. Through his telescope Galileo saw that the Milky Way was not just a giant glowing cloud, but a vast field of stars like our own sun. Galileo saw what no one else had seen before him, that the stars and their planets were suns and worlds, like ours. He saw that the universe was imbued with change and was not an immutable creation as had been envisioned.

## BIRTH IMPLIES DEATH

It's ironic that Galileo made so many of his discoveries about the universe in Venice, Italy. During Galileo's lifetime, even though people realized that the Earth changed, and nothing built by man would last forever, most people believed that at least the stars and the heavens were permanent. Galileo's observations changed that view forever; his findings proved that the rest of the universe was not immutable and was in a constant state of change. Even the stars are born and die. Since the time of Galileo, scientists have used telescopes to look to the very ends of the universe, and have used particle accelerators to peer into the hearts of atoms, and they have found change everywhere they've looked. We now know that Venice, itself, where Galileo made his discoveries, is sinking deeper into the water with every winter storm and spring rain; it cannot last forever. Venice is a very good example of the old notion that birth implies death. Like the city of Venice, we are all destined to die. We will not live forever, but we are not alone in that – it appears that neither will the planets and the stars live forever, nor the atoms that make them up.

Venice Italy, the home of Galileo. It is here that Galileo's observations married the Earth to the rest of the Universe.

# CHAPTER 9: THE TRAIL TO THE BEGINNING OF TIME

If the electromagnetic force and weak force combine at high energy, and if the electro-magnetic, weak and strong forces combine at even higher energy, then perhaps all four forces would combine at an energy level that was higher still. There is no lab on Earth that can create the energy level needed to combine all of the four forces into a single, unified force. To see the four forces unite into one, we must look back to the beginning of time, to the creation of the universe itself, at what scientists refer to as the Big Bang.

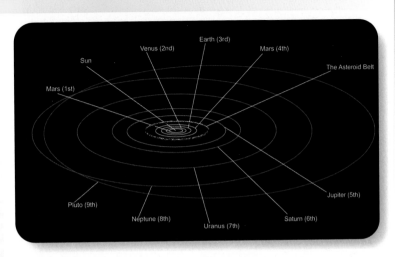

Earth is the third planet in our solar system of nine planets which orbit a star, our sun, also called Sol.

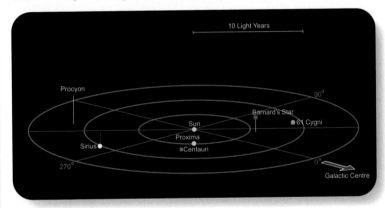

Our solar system is surrounded by thousands of stars of various kinds, each of which may have planets with life orbiting them.

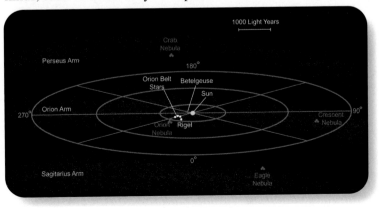

This is our home, the Earth, as viewed from outer space.

Our solar system and all of our close neighbor stars reside within one of the many spiral arms of the Milky Way galaxy, a huge system of more than 250 billion stars.

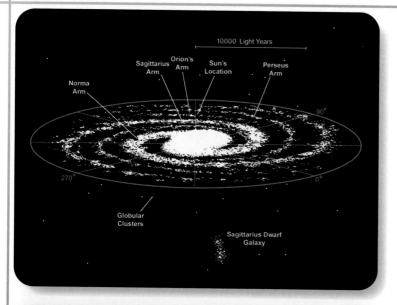

Here is our home galaxy, the Milky Way, a mid-sized galaxy that is orbited by two small satellite galaxies and a score of globular clusters. Globular clusters are large collections of dozens or even hundreds of stars that are tightly packed together.

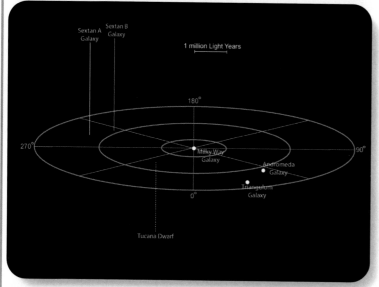

This is the local group of galaxies, of which the Milky Way is a member. The galaxies that make up the local group are all within a few million light-years of Earth. Some two million light-years away from our galaxy sits the Andromeda galaxy. Andromeda is a spiral galaxy, like ours, and appears to have a very similar structure to the Milky Way.

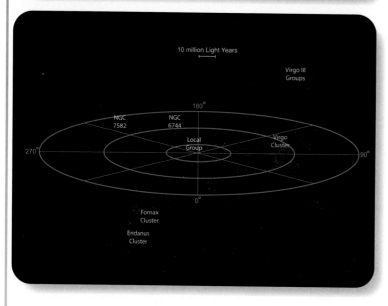

The local group is part of a huge archipelago of galaxies known as the Virgo Super Cluster. The Virgo Super Cluster expands over tens of millions of light-years and is the home of thousands of galaxies. Careful observation of the cluster has revealed that the universe is expanding. Astronomers believe that at the rate the universe is expanding the Virgo Super Cluster should break apart into much smaller clusters of galaxies within the next 50 billion years.

The Virgo Super Cluster is itself a part of a huge collection of "local" super clusters, which spreads over hundreds of millions of light-years. Astronomers have identified dozens of super clusters that make up the local super cluster group.

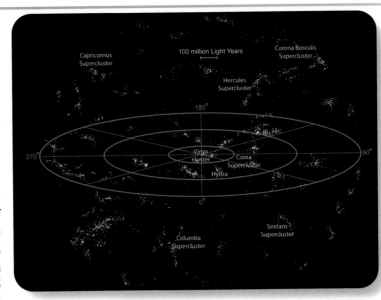

The local super cluster groups are part of huge filaments of galaxies that stretch across the entire universe. Cosmologists estimate that the known universe stretches over 15 billion light-years and is expanding in all directions. If the universe is expanding, then cosmologists believe that many billions of years ago all of the matter – galaxies, stars and gas – in the universe may have been contained in a single, infinitely small point. If we were to follow the expansion that we see in the universe today backward through time, towards that single, infinitely small point, we would find that the ambient energy level in the universe was continually increasing, which leads the cosmologists to believe that exactly the kind of environment that physicists maintain would allow for the unification of the four forces did, in fact, exist in the early stages of our universe.

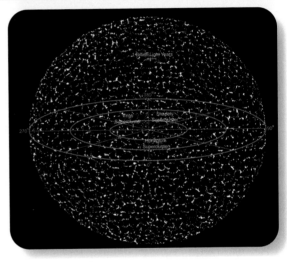

Albert Einstein is regarded as one of the most brilliant people ever to have walked the Earth. He is famous for, among other things, his two theories of relativity: special relativity and general relativity. Special relativity is a simplified scenario which ignores gravity, while general relativity, which came later, is much more comprehensive. It is Einstein's special relativity theory that asserts that nothing in our universe can be accelerated to the speed of light. The general theory of relativity was first published in 1915; this theory predicted, among other things, that the universe was expanding in all directions. Einstein himself did not accept that conclusion and restructured his theory so that his theoretical universe would stand still. He eventually reversed himself and accepted an expanding universe, calling his earlier reluctance "the worst blunder of my career."

While Einstein was trying to restructure his own theory, the astronomer Edwin Hubble, for which the Hubble Space Telescope is named, was peering into the universe, trying to probe its secrets. Without knowing of the predictions of the universe made by Einstein's general theory of relativity, Hubble found through telescopic observation what Einstein had predicted – that the universe was expanding. Einstein and Hubble met for the first time in 1931 and compared their theories and observations. They concluded that Einstein had been correct from the start.

If the unified field theories are correct, nothing in the universe will last forever. The stars, galaxies, and even the atoms that make them up, will some day all decay and die.

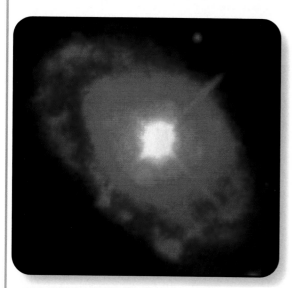

## GIANT QUASARS

Using giant ground-based telescopes and the Hubble Space Telescope astronomers have been able to see objects deep in space, and therefore back in time. Here is a false color, computer enhanced picture of a quasar. Quasars are thought to be energetic galaxies with massive black holes at their centers, going through a fairly violent youthful phase. Most quasars can be found 10 to 12 billion light-years out into space. This means we are seeing these objects as they appeared 10-12 billion years ago. It took their light that long to reach us, even traveling at the fantastic rate of 186,000 miles per second. The quasars appear to be nearly at the limits of what we can see in the expanding universe.

Einstein and Hubble were able to see something that no one before them had seen – that the universe was expanding. The expansion of the universe is not as simple a concept as it at first seems. One of the more frequently asked questions is: "if the universe is expanding, then where is the center of the expansion?" The answer – there is no center of the expansion. A simple analogy can be made by using rubber balloons. If you draw dots on the surface of a balloon and then blow it up, all of the dots appear to be racing away from each other. If you place yourself on any one of the dots, then, it would appear that all of the other dots are racing away from you. Likewise, if you take all of the air out of the balloon then it would appear that, regardless of which dot you place yourself on, all of the other dots are moving towards you. From this perspective, every point in the universe can be seen as the center of expansion, and space itself, between stars and galaxies, is expanding right along with the matter that it contains.

Here is a computer simulation showing the locations of all of the known galaxies in our universe. The darkened triangular areas represent regions for which no data is available.

If we look deep enough into the universe, we can see several quasars at the edge of known space. Beyond these points, astronomers don't currently see anything. It is believed that this is because they are looking at points so far distant that, since they are seeing conditions as they were so far in the past, the stars, galaxies, etc. have not yet progressed to the point where they are organized into structures of the familiar matter as we see today. Cosmologists refer to this time as the epoch of darkness. Using sophisticated radio telescopes, observers can see another form of radiation that appeared subsequent to the Big Bang. This "background" radiation can be detected all over the universe, but it has thinned out so much as the universe has expanded that it very early in time dropped from the visual range (light) to radio frequencies. Radio telescopes are not the only way to see this radiation. It can be seen using a fairly common device, that we are all familiar with, a television set. If you have a television set with an antenna, and tune your TV to a dead channel, you can see this radiation. Turn down the color on your TV until all you see is a black screen dotted by flecks of snow. A small percentage of those flecks are actually photons, carriers of the electromagnetic force – relics from the creation of the universe that have been speeding through space since the beginning of time.

The legacy of creation is still with us. The energy released by the sun and other stars, from the fusion of their atoms, is only a fraction of the total energy that has been with us since the beginning of time. It was only at the beginning of time, in the brief period immediately following the Big Bang, that the universe could have worked under the simple laws predicted by the unified field theories.

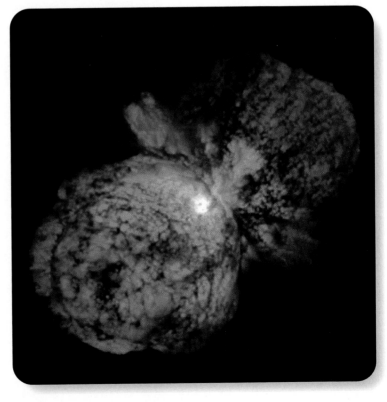

# CHAPTER 10: IN THE BEGINNING

Looking at nature on the very small scale, we see that its basic structures and the forces that govern it would be simpler under conditions of extremely high energy and temperature, what we call the equilibrium thermal energy. By looking out into the universe and observing the farthest galaxies, we see that the early universe functioned under these conditions of high energy and temperature. By putting these two realms of investigation together, physicists and cosmologists have been able to trace the broad outlines of the history of the universe, from the first fraction of a second following creation to the present day. Now, scientists don't know everything by any means; many of the theories that they've put forth are inaccurate, contested, missing pieces, and even flat-out wrong. But despite this, scientists do have a very good understanding of what the early universe was like. In this chapter we will examine the scientific account of the creation of the universe and examine in detail the key events that helped to form the universe that we know today. It is the merging of cosmology and astronomy with atomic and subatomic physics and quantum theory that gives us an understanding of how the universe around us came to be. The following is a brief account of what scientists know about the formation of the universe. Each picture will give you an account of an early epoch in the history of the universe, moving back towards the first instant of creation.

## THE EARLY UNIVERSE: 3 BILLION YEARS AFTER THE BIG BANG

Here is our galaxy, the Milky Way, as it looked about 12 billion years ago. This is a snapshot of our home galaxy only a few billion years after the creation of the universe. Our sun has not yet formed and the vast majority of the galaxy was made up of only two gases – hydrogen and helium. The periodic table of elements that we find at the front of every chemistry class would be remarkably simple.

## THE ERA OF THE QUASAR: 2 BILLION YEARS AFTER THE BIG BANG

During the first one to two billion years after the universe came into being, it was dominated by quasars. Quasars are highly energetic galaxies that produce a vast amount of electromagnetic radiation, some of it as light. Quasars can be seen all across the universe, right out to the boundaries of what we can observe, and a single quasar can have a luminosity equivalent to thousands of normal galaxies. Above is a NASA photo of a quasar billions of light-years away in space.

## THE EPOCH OF LIGHT: 1 MILLION YEARS AFTER THE BIG BANG

By approximately one million years after the Big Bang, the universe had expanded and thinned out enough that photons, the carriers of the electromagnetic force, could move freely through space without constantly bumping into any other particles. Cosmologists refer to this time period as the Epoch of Light. This was also the era of the first atoms. Free from the pestering of photons, electrons were now able to settle into orbit around protons and neutrons and form the first atoms. The most common atoms to form, indeed the most common atoms in the universe, were hydrogen and helium. Hydrogen is the simplest of the elements, consisting of one electron orbiting a single proton. Helium has a nucleus with two protons and two neutrons being orbited by a pair of electrons. Although the helium atoms formed during this period, their nuclei (cores), made up of protons and neutrons, were formed earlier in the life of the universe.

# THE FORMATION OF ATOMIC NUCLEI: 100 SECONDS AFTER THE BIG BANG

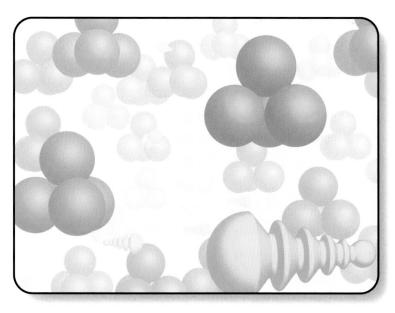

Helium Nuclei were created at roughly 100 seconds after the Big Bang. By this time, the universe had cooled down enough for protons and neutrons to get together and form the helium nuclei, bonded together by the strong force. Before this era, the ambient energy of in the still cooling universe would cause protons and neutrons to be forced apart if they attempted to form nuclei. In this early epoch, protons and neutrons tended to form groups of threes – two Protons combining with a single neutron. A pair of these triplets, in turn, tended to bond together into a larger nucleus with four protons and two neutrons. Two protons were easily thrown off of this larger nucleus, resulting in a stable helium nucleus (two protons and two neutrons). Physicists estimate that this early helium nuclei dance resulted in a universe that was 25% composed of helium. Cosmologists have done elaborate experiments and confirmed that about ¼ of the universe is in fact composed of helium.

# THE QUARK FOG: 1 SECOND AFTER THE BIG BANG

At one second after the Big Bang, the heat energy of the universe was so intense that the strong force could not function. Protons and neutrons could not form because the strong force was not available to keep them together. Quarks, unable to combine into protons and neutrons, existed freely – the universe was a quark fog mixed with photons.

## NEAR INFINITE DENSITY: $1/10^{th}$ OF A SECOND AFTER THE BIG BANG

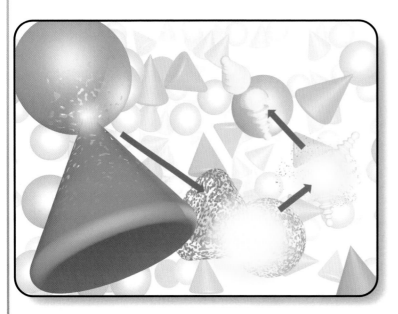

At this point in history, the universe was so dense that even neutrinos found it hard to move. Neutrinos are sub-atomic particles so small that they can pass through thousands of miles of solid lead without touching anything!

## THE ELECTROWEAK ERA: $1/10$ BILLIONTH OF A SECOND AFTER THE BIG BANG

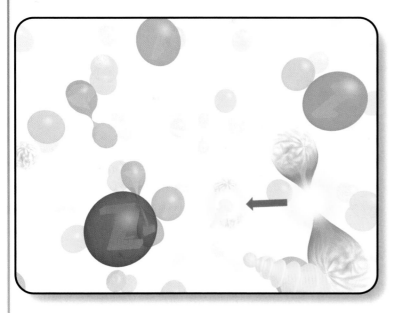

There was enough ambient energy at this point in the universe for the electro-magnetic force and weak force to join into the electroweak force, carried by both photons (electromagnetic) and weak bosons (weak force) exchange particles, which now exhibited similar properties. These exchange particles could be created in an overwhelming supply. It was here that the weak bosons and photons moved interchangeably and three forces, not our current four, governed the universe.

# THE GREAT BATTLE, MATTER VS. ANTIMATTER: $10^{-32}$ SECONDS AFTER THE BIG BANG

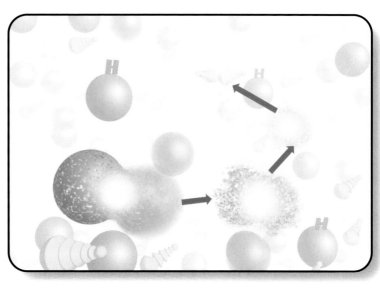

The universe is now in the grips of the ultimate battle. Quarks, the fundamental building blocks of all matter, are now in a titanic war against antimatter quarks. When the matter quarks meet their antimatter counterparts, the result is mutual annihilation. However, there was a small imbalance in the amounts of matter and antimatter in the early universe. For every billion anti-quarks there were a billion and one normal quarks. At the end of the matter / antimatter battle, a small amount of normal matter remained. This matter became everything that we see in the universe today. The energy left over from that battle can still be observed as the cosmic background radiation. At this point, Higgs bosons, hypothetical particles that give the mass to matter, are created out of the energy that drove the universe through its early inflation period.

# $10^{-35}$ SECONDS AFTER THE BIG BANG

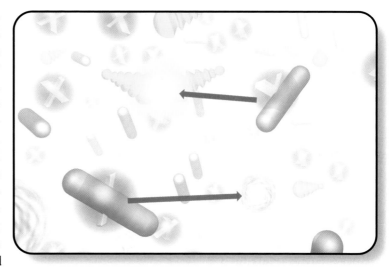

During this brief instant in history, the strong force had not yet broken away from the electroweak force, and only two forces ruled the universe, instead of three or four. This was the era of the electro-nuclear force; for one brief, glorious moment the unified force envisioned by physicists and cosmologists (the unification of the electro-magnetic, weak and strong forces) was a reality. The theoretical X particles traveled through the universe as the exchange particle for the electromagnetic, weak and strong forces – at this level of ambient thermal energy, gluons, photons and weak bosons were interchangeable. So, at this juncture there were only two forces controlling the fate of the universe – the electro-nuclear force (or grand unified force, as it is sometimes called) and the force of gravity. This is the beginning of the era that cosmologists call the inflation period. Immediately following this moment in time the strong force broke away from the electroweak force. This breaking away was believed to cause a super-rapid expansion of the universe, and this super-expansion kept the newly formed universe from collapsing back in on itself under the effects of gravity.

# $10^{-46}$ SECONDS AFTER THE BIG BANG

We have now reached the first instant of creation. There is no name yet given to time increments as tiny as those between the first events of creation. Mathematicians symbolize the time to this first milestone by a decimal point followed by 46 zeros. The universe at this point was a potent kernel of energy, rapidly expanding, yet smaller than today's smallest atom. The universe was ruled by one primordial law that combined all four forces into one "superforce." Electromag-

netism, the weak force, the strong force and gravitation all behaved in accordance with the characteristics of this one superforce, which could explain the behavior of the entire structure of the universe – i.e., a Theory of Everything. If we could know how the universe behaved at this point in its history, then we might finally understand the exact nature of our universe, and the relationships between matter and energy, and time and space. Finally, we might be able to derive the unified field theory that so confounded Einstein. But right now, we don't know.

We lack a theory, at present, to explain the early universe completely, and how it would have behaved under these extreme conditions. Many physicists and scientists are looking for a more complete equation or theory; some believe it will be a sort of supersymmetry, Superstring theory, quantum gravity, or some other equally fancy theory. For now, although we don't know exactly what this theory will say, where it might lead us, we do know that whoever finds it will be the first person in history to understand how the universe came to be, and quite possibly, the first, as Einstein said, "to understand the mind of God."

## THE UNIFICATION SCALE

Here is a simple diagram of how the four forces in nature combine as we move closer to the beginning of time (the Big Bang). Many hope that what is called a "Theory of Everything" will allow us to understand how all of these forces combine into a single "superforce," and tell us how the universe works.

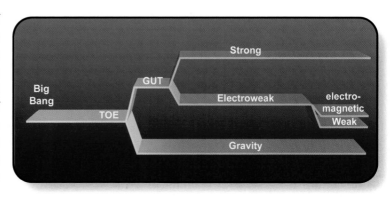

## BROKEN SYMMETRY

This sculpture, on the main road that leads to Fermi Labs just outside of Chicago, is a fitting tribute to what the men and women at Fermi Lab are in search of – a way to unbreak the broken symmetry of the universe. What physicists call broken symmetry was absolutely essential for the universe, as we know it, to exist. As we look deeper into the past, we see more and more evidence that the universe began in a state of perfect symmetry – a state wherein all forces, all matter, space and time were all one perfect entity, obeying one perfect law. The breaking of that perfect symmetry as the universe expanded and cooled allowed for the creation of forces, matter and energy as they exist today, resulting in the creation of everything we see in our universe. We owe our very existence to broken symmetry.

# CHAPTER 11: GRAVITATION; THE HOLDOUT FORCE
## PART 1 – NEWTON'S LAWS OF GRAVITY

Of the four forces in nature, gravity appears to be the most stubborn. Unified field theories state that only at unbelievably high temperatures could gravity merge with the other three forces. Gravity is somewhat of an enigma and has been "rewritten" on three different occasions in attempts to explain its principles correctly. Each time the theory of gravity has been rewritten, it has caused both elation and frustration to the world of science. We owe our present understanding of gravity to two men: Sir Isaac Newton and Albert Einstein. Although we have tweaked our understanding of gravity since the times of each of these two men, their work provides the basic framework for understanding gravity and its relation to the rest of the universe.

In 1687 Isaac Newton brought forth a series of theories and mathematical concepts to explain how gravity worked. It is rumored that Newton's inspiration came from watching an apple fall from a tree in the gardens at the world famous Woolsthorpe Manor. Newton reasoned (after watching the apple fall) that the apple fell because of a force of attraction, which he called gravity. He decided that the more massive (heavier) an object was, the greater was the attraction that it exerted on other objects, i.e., its pull of gravity. Therefore, the apple fell to the Earth, and not the other way around.

## THE START OF SOMETHING BIG

After watching an apple fall from a tree in Woolsthorpe Manor, Isaac Newton pondered why all things had a tendency to fall back to the Earth. This line of inquiry would lead Newton to create a series of laws and theories detailing the workings of the force that he called gravity.

## DISTANCE IS KEY

Newton's theory of gravity states the farther apart two objects are, the weaker the gravitational attraction between them is. His theory also states that the strength of gravity decreases with the square of the distance. If you double the distance between two objects, then they feel only one quarter of the gravitational attraction. This theory can be tested on Earth. An object at the bottom of a tall skyscraper weighs slightly more than it does at the top of the skyscraper.

## THE MOON IN ORBIT

The Moon orbits the Earth because of their mutual attraction – gravity keeps the two worlds together. If the Earth were to suddenly disappear, then the Moon would go flying off into space in a straight line. The Earth's gravitational pull keeps the Moon from leaving home, and keeps it in orbit.

Newton's laws of gravity tell us that it also increases in proportion to the masses of the objects involved. To break the force of gravity that holds you to a massive object, like the Earth, you must apply a force (stronger than the gravitational force) away from the Earth (in the opposite direction to gravity's attraction). Unless you continue to apply that force, gravity will pull you back down again. If you want to "get away" from the Earth, there are only two ways to do it: (1) continue to apply a force

greater than gravity until you're far enough away that Earth's pull on you is negligible; or (2) get traveling away from the Earth very rapidly, specifically, faster than what is called "escape velocity." The escape velocity from Earth is 11 kilometers per second (which is about 25,000 miles per hour!). The Moon, having much less mass than the Earth, has an escape velocity of only 1.8 kilometers per second.

The more massive an object is, the greater its gravitational pull on other objects, and the larger its escape velocity. The sun's escape velocity is 620 kilometers per second. There are objects in the universe, called black holes, that are so massive that escape from them is not possible – their escape velocity is, in fact, greater than the speed of light (186,000 kilometers per second), which is why they are "black." We know of nothing that can travel faster than the speed of light, so anything that falls into a black hole can't possibly escape and is lost forever.

## BUT WHAT IS IT?

Newton established general laws to explain how gravity works, but was himself unsure as to exactly what gravity actually was. This was similar to feeling the wind on your face, and knowing that it could drive a windmill, but not knowing what the wind was made of. The world would have to wait until 250 years after Newton's time for someone to explain more fully what gravity is.

# CHAPTER 12: GRAVITATION; THE HOLDOUT FORCE
## PART 2 – EINSTEIN'S GENERAL RELATIVITY

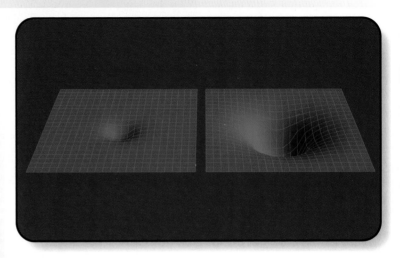

For more than 250 years Newton's laws of gravity ruled supreme, until Albert Einstein put forth his general theory of relativity in 1915. Einstein went one step further than Newton, who saw gravity only as a force. Einstein saw gravity as a distortion of space and time. Although Newton's theories of gravity are an accurate account of how gravity behaves in everyday situations on Earth, Einstein's general relativity is far more accurate, and explains the "what" as well as the "how" of gravity.

## WARPED SPACE

Einstein saw gravity as the warping of space. If the universe is visualized as a flexible two-dimensional membrane (like Saran Wrap stretched over a bowl) and you place a heavy object, like a baseball, on the Saran Wrap, you get a little indentation. This indentation, or "warping" of the Saran Wrap, is analogous to the way that gravity "warps" space. As Einstein envisioned it, the greater the mass (planet), the greater the indentation (warping of space), and the greater the "gravitational attraction" that results. Taking our analogy a step further, if you place a marble on the Saran Wrap, it will immediately roll towards the indentation made by the baseball, stopping only when it is in contact with the baseball. Similarly, in the real world, if you (small mass) were placed near the Earth (large mass) but not touching it, gravitational attraction would immediately accelerate you towards the Earth until you came into contact.

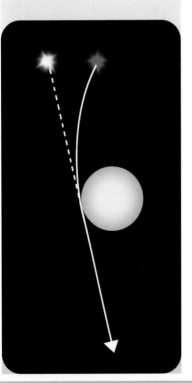

## GENERAL RELATIVITY IN ACTION

Einstein theorized that since strong gravitational fields warp space, then a beam of light traveling near a strong gravitational field should be deflected from its straight line path. Many scientists couldn't accept this part of Einstein's theory, but he stood by his theory and even went as far as to say that the sun's gravity should bend light passing behind it. Astronomers finally tested his theory in 1919, during a total solar eclipse of the sun, the only time when stars close to the sun can be seen. The apparent position of the stars shifted slightly exactly as Einstein had predicted years earlier.

Einstein's predictions from general relativity state that the more massive an object is, the more space becomes distorted by it, and the more powerful its gravitational field (i.e., the deeper the gravity well). This is what Newton didn't understand about the nature of gravity. Einstein realized that gravity was merely the distortion of space caused by the mass of an object. According to Einstein's theory, any object that has mass will distort the fabric of space, no matter how small it is.

Below is a series of increasingly massive objects placed on a two-dimensional membrane; we see that the more massive an object is, the greater the distortion it creates. It's important to note that it is mass, not physical size, that determines the amount of distortion. As we'll see further on, some things in our universe are so dense that they have enormous mass for their size.

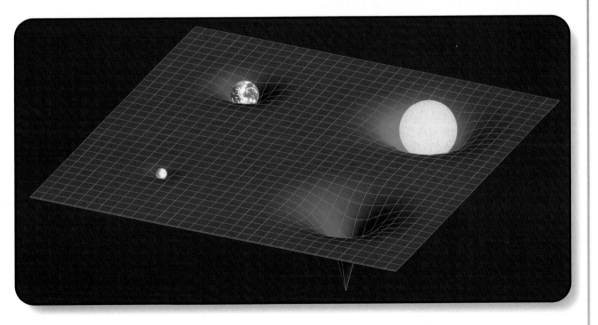

## EINSTEIN'S MOST REMARKABLE PREDICTION

Einstein's general theory of relativity states that objects can exist in our universe that are so massive their gravity becomes overwhelmingly strong, so strong that nothing, not even light, can escape its pull. If the massless photons of light, the fastest thing in the universe, cannot escape the gravity of these objects, then nothing can escape them. Astronomers have dubbed these bizarre objects "black holes," because they act like holes in space, trapping anything that falls into them, and since light cannot reflect from them, they appear completely black – only by their effects on matter in their vicinity have astronomers been able to locate them. Above is a rendering of what a black hole may look like as it consumes energy and matter from a nearby star. Despite the fact that these objects were predicted by his general theory, Einstein himself refused to believe in these objects. Black holes are an example of gravity gone mad.

## THE GRAVITON

Newton made the first big leap towards understanding the force of gravity; Einstein made the second leap with his general theory of relativity. The third important leap came with the unified field theory and quantum theory (the branch of physics that explains the world of the very small), which claims that gravity is more than just a force and a bending of space. According to these theories, gravity is carried by an ultra-small exchange particle known as the graviton. No one has yet detected a graviton, but in order for unified field theories and TOE (the Theory of Everything) to work, gravitons must be found. If gravitons can be experimentally detected, then perhaps it will lead the way to a Theory of Everything.

# CHAPTER 13: BLACK HOLES AND THE THEORY OF EVERYTHING

Einstein's theory of general relativity says that the gravitational field of a black hole is so massive that nothing, not even light, can escape from it. A black hole is the remnant of a massive star that has collapsed in on itself once the expanding force of the nuclear fires at its core was overcome by the contracting force of its gravity. At the center of a black hole is a singularity. The singularity is the monster that powers a black hole; it is, for all intents and purposes, a point of infinite density that occupies no space. Black hole singularities have properties similar to the infinitely dense point from which cosmologists believe our universe was formed. Therefore, many physicists postulate that an examination of the singularities found within black holes may yield clues to the origin and possible evolution of our universe, and possibly lead to clues which point towards finding a Theory of Everything.

## THE FORMATION OF A BLACK HOLE

A singularity is found within the heart of a black hole. Black holes form at the death of super-massive stars when they collapse and form super-dense objects that nothing, not even light, can escape from. On the opposite page is the step-by-step progression in the formation of a singularity.

# A BIRTH IN GAS AND DUST!

All stars are born from giant gas clouds like this one. The gas and dust begin to condense under the pressure of their mutual gravity. Most of these clouds are composed chiefly of helium and hydrogen, the most abundant elements in the universe. As gravity continues to have its way, the cloud of gas and dust eventually collapses into a solid mass, such as a planet. However, if the total mass is sufficiently large, it has the makings of a new star.

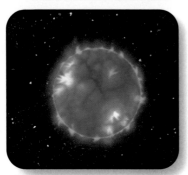

# BRIGHT YOUNG STAR

When a massive enough cloud condenses, conditions are right for nuclear fusion to begin, i.e., we have ignition, and a star is born. The star's fusion reaction is the conversion of hydrogen to helium, but in larger stars (like our sun), once all of the star's hydrogen "fuel" has been consumed, it starts fusing the helium into carbon and oxygen. In very massive stars (more than five times as massive as our sun), the carbon and oxygen can subsequently be fused into even larger atoms. Young stars are usually white hot or blue hot stars. Our sun, more than likely, looked like this during its youth.

# THE MIDDLE AGES

During a star's "middle age" it undergoes a slow and steady expansion, gradually getting bigger and hotter. At this point the star has been shining brightly for some 5 billion years (roughly the present age of our sun).

# THE RED GIANTS

Eventually, as our sun runs out of fuel, it will expand and devour our inner solar system. Mercury, Venus, Earth and Mars will be boiled away into nothingness. The sun will have become a red giant, blowing its outer layers off into space as it expands. Eventually, after approximately another billion years, our sun will start to recondense, beginning the slow process that will lead to its death.

# DEATH AS A WHITE CORPSE

As a star blows off more of its layers into space, it will shrink. Once it has used up all of its nuclear fuel, all that's left is a shrunken white core, called a "white dwarf," and is not much larger than the size of the Earth. The material that makes up the white dwarf is so dense that a teaspoon of it would weigh as much as an elephant. This is how most lightweight and mid-sized stars will die. But a very bizarre end is in store for the super-massive stars.

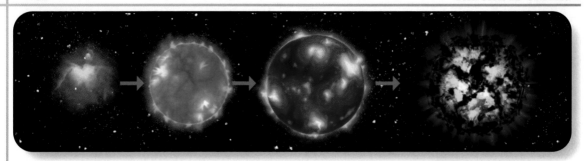

## GOING OUT WITH A BANG!

A super-massive star (with a mass more than ten times that of our sun) exhausts its fuel reserves within a few million years, instead of billions. This kind of star is so massive, and expends fuel at such a furious rate, that it forms an iron core at its center. There is not enough (fuel) energy in any star to cause the fusion of iron, so once all of the lighter elements have been consumed, the infalling (gravitational) energy of the star brings all of its mass crashing down onto the iron core, where it violently rebounds and blows the star apart. The result is . . .

## A SUPERNOVA!

The massive star literally explodes! The outer layers are blown off into space in a titanic super-explosion. It is one of the most violent events in the universe. But while the outer layers are blown away, the core collapses in on itself. The core continues to collapse until all that's left is a super-dense star the size of New York City. This star is so dense that all of the protons and electrons in the atoms of the star are "squeezed together" into neutrons which, along with the neutrons already present, leaves a star that is made almost solely of neutrons. For obvious reasons, astronomers call them neutron stars.

## NEUTRON STARS

A neutron star is the end of the road for many of the most massive stars. They are all that is left of the massive monsters from which they formed. Neutron stars are so massive and so dense that a few grains of material from a neutron star can weigh as much as an aircraft carrier.

## PULSARS

Some neutron stars spin very rapidly (some have been known to spin upwards of 600 times a second). As the neutron star spins, it throws off intense radiation along its magnetic axis. Twice per rotation (at the opposite poles) the star's magnetic axis points in our direction, and we can detect its brief passage here on Earth. Scientists have dubbed these spinning neutron stars pulsars because, from Earth, they appear to be pulsing on and off.

## BLACK HOLES!

Extremely massive stars may end their lives as even more bizarre objects than neutron stars. Einstein's theory of relativity tells us that if a star is massive enough then, after it blows off its outer layers in a supernova explosion, its core will collapse down to something even smaller than a neutron star. In fact, physics, as we currently understand it, states that nothing can stop this collapse and the star will continue to collapse until it reaches a condition of infinite density and becomes a point in space. Cosmologists call this object a black hole. The gravity of a black hole is so powerful that anything falling into it is doomed. Nothing, including light, can escape from it. Black Holes are gravity gone mad! Below is a computer rendition of matter and energy falling into a spinning black hole. The contents of nearby space swirl around the perimeter of the black hole like a cosmic whirlpool until gravity pulls them into the black hole, never to be seen again.

## LOOKING FOR BLACK HOLES

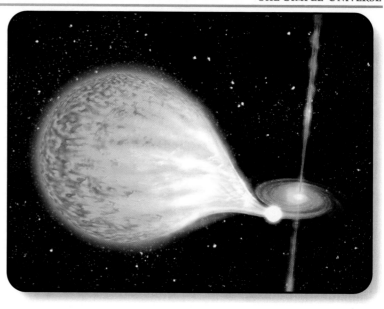

No one has ever seen a black hole – their very nature keeps them hidden. We can only detect an object when light, or other detectable energy, is reflected from it; since a black hole absorbs rather than reflects all energy, it remains invisible. However, its effects can be seen on nearby matter as that matter swirls toward and falls into the black hole. An analogy can be seen on Earth with tornadoes and twisters: you can't see the wind in a tornado, but you can see the material and debris that it whips around. In the computer image to the right, although the black hole is not actually seen, its effect on the nearby giant star can be as it strips matter away from the star and swirls it about in a giant gravitational whirlpool before devouring it.

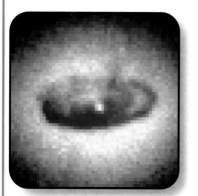

## BLACK HOLE AT THE GALACTIC CORE

At the center of the galaxy named NGC-4261 a super-massive black hole has been detected that is devouring everything within its reach, including stars and giant gas clouds. Many suspect that at the center of our own galaxy lies a black hole as massive as the one in NGC-4261.

The singularity (the super-condensed, super-massive core of the dead star) is what powers a black hole. Physics, as we know it, tells us that within a singularity the laws of nature and the laws of relativity break down. This is exactly what most scientists believe happens as we look at conditions closer to the beginning of time. As we move backward through time toward the Big Bang, the behavior of matter and energy begin to change, the laws of contemporary physics no longer apply, and the four individual forces begin to merge into a single superforce.

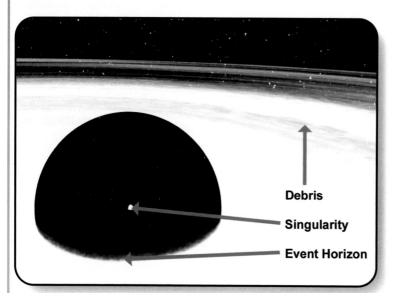

Debris

Singularity

Event Horizon

## UNIVERSE SIZED BLACK HOLE

As mentioned previously, the general theory of relativity tells us that our universe is expanding and that perhaps, one day, it will reach a maximum level of expansion, and then begin to contract back in on itself. The universe has no mathematical center of expansion; however, every point in the universe, from its own perspective, appears as the center of expansion – no matter where you are, everything else in the universe appears to be moving away from you. Likewise, if the universe were to begin to contract, no matter where you were, you would seem to be at the center of the contraction. This is very similar in context to being within the influence of a black hole, where it is impossible to escape – no matter how far or how fast you travel in our universe, you are at the center of the expansion (or collapse) and can never escape. So, its as if we were all inside a universe-sized black hole.

## THE COLLAPSE OF PHYSICS WITHIN A SINGULARITY!

The study of singularities may hold the key to finding our Theory of Everything, and help to define a unified field equation that would explain how the four forces unite in a very high energy environment. At the very least, a study of singularities may yield clues to the future of our universe, and how the laws of physics may mutate in the millennia to come if expansion reaches a maximum limit and universal contraction follows.

Einstein's theory of relativity states that around and within a singularity, the conventional laws of physics break down. Cosmologists believe that the closer we get to the center of a black hole, the further the laws of physics deviate from what we know, just as Einstein predicted. Likewise, theories about the creation of the universe include the startling prediction that our universe began as the ultimate singularity. Given the apparent similarities in the realm of black hole singularities and the early stages of our universe, study of black holes may help cosmologists understand the singularity that gave birth to the universe.

# CHAPTER 14: SUPERSTRING THEORY AND THE THEORY OF EVERYTHING PART 1 – A THING FOR STRINGS

In the 20th Century, there were two major advances in the field of physics that gave science a blueprint for the underlying simplicity and elegant laws that govern our universe. These two pivotal concepts are general relativity and quantum theory (also called quantum physics). Quantum theory provides an understanding of the interplay between the electromagnetic, strong and weak forces in the realm of the subatomic. General relativity explains the gravitational force in the terms of the geometry of space and time. Each is a powerful predictive tool in its own right, but all past attempts to bring the two together have ended up in conflict. In recent years, however, a new field of mathematical physics called superstring theory has been developed. Superstring theory, being fairly new, is not yet completely understood, but early results seems to hint that perhaps superstring theory may finally be able to tie together quantum theory and general relativity. Some physicists are so excited about this new field of physics that they believe we are finally close to finding a unified field theory – our Theory of Everything.

## A STRING OR A POINT?

Quantum physics (the physics of the subatomic world) views a fundamental particle as equivalent to a mathematical point in space, flashing into and out of existence. As they flicker in and out, these points convey the electromagnetic, weak and strong forces. Mathematically, these force carriers are considered to be dimensionless. Physicists have always assumed that gravity could also be described in this manner, being carried by a point-like particle. However, when general relativity (Einstein's theory that includes how gravity behaves) is applied to the quantum world, it breaks down. Instead of giving an accurate view of how gravity behaves on the subatomic level, the laws of relativity appear to become lost in a situation where all four fundamental forces have more equal effects.

In the last 25 years, physicists have developed a new theory which may eventually lead to a truly unified field equation or Theory of Everything. Mathematicians and cosmologists have accidentally discovered that by replacing the dimensionless point particles in their thinking with one-dimensional strings – with length, but no thickness – all of the mathematical problems associated with gravity at the quantum level seem to disappear. Mathematicians have proposed that there are two types of strings: open strings, with free ends; and closed strings, like a loop. Both types of strings are extremely small. To be exact, these strings are 0.000000000000000000000000000000001 cm ($10^{-33}$ cm) in length! On a more practical scale, if a regular atom were the size of our solar system, then a string would be the size of an atom! Most physicists tend to believe in closed string theories, as opposed to open ones, because closed strings seem to generate smoother, more realistic equations.

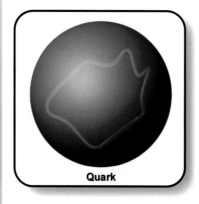

**Quark**

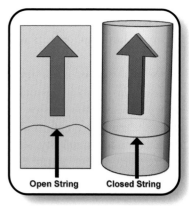

**Open String**    **Closed String**

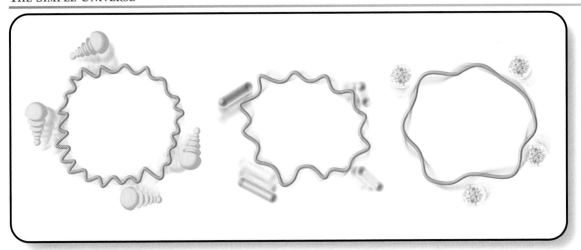

## SOUNDS OF MUSIC!

Superstring theory (SST), as currently understood, tells us that the vibration of a string can produce a variety of particles. All of the particles that we see in our universe today may be the manifestations of string vibrations and the properties inherent to them. This can be envisioned in simple terms by a guitar string: varying the tension of the string (tuning it higher or lower) produces different musical notes when plucked (caused to vibrate). When a subatomic string vibrates, it produces particles instead of notes. The higher the energy state at which a string vibrates, the greater the mass of the particle that it produces. As strings vibrate they are capable of producing all sorts of particles. These particles can theoretically range from the smallest point particle to a particle the size of a mote of dust, if not larger. Physicists have never found any particles the size of dust motes (or anything even remotely close for that matter). Strings have properties to characterize them, such as spin, vibrational intensity and frequency.

## STRINGS INTERPLAY

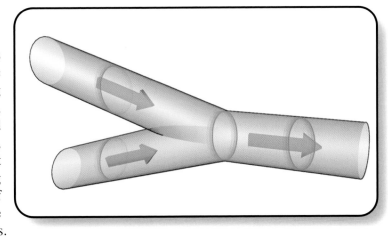

Classical quantum physics dictates that subatomic particles exchange forces using force carrier particles (photons, weak bosons, gluons and gravitons). String theory, however, takes a different view. According to string theorists, the exchange of forces can be much more easily explained using strings. String theory states that an exchange of forces is actually the merger of two strings and the resultant sharing of their energy.

Depending on the excitation state of a string, many types of force exchange can be explained. This complex and constant merging and separating of strings can help to explain the complex topography of the quantum universe.

# CHAPTER 14: SUPERSTRING THEORY AND THE THEORY OF EVERYTHING PART 2 – A TALE OF A STRING

Strings are immensely small; because of their micro-subatomic size, they will be very difficult to detect. Physicists calculate that a string would be $1/100,000,000,000,000,000,000$ the size of a proton, and that is mighty small. Things of this size are measured in the Planck scale (where the "Planck length" = $1.6 \times 10^{-33}$ cm). This scale is used to discuss the smallest known things in our universe. String theory has led to a series of bizarre ideas about the unseen universe. Among them are theories that our universe may be comprised of not just four dimensions (time plus three spatial dimensions), but perhaps as many as 9 to 12 dimensions! Strings may wrap themselves into a collection of unseen dimensions, completely undetectable on the macro scale at which we live.

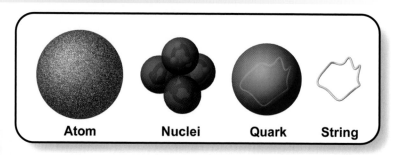

**Atom          Nuclei          Quark          String**

## THE STRING IS THE THING!

Conventional physics teaches us that all matter in the universe is made up of atoms. These atoms in turn are made up of very small elementary particles. String theory takes things a step further, asserting these elementary particles are themselves made up of ultra, ultra small strings, which exist at the tiniest levels possible. The vibrations of these strings, according to theory, are the root cause of everything in the universe. These strings may inhabit unseen spatial dimensions, and be the controlling mechanism of all events since the beginning of time.

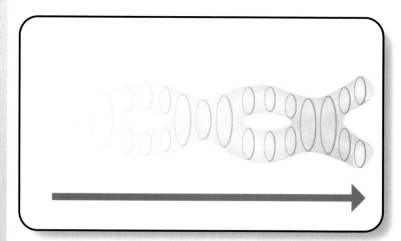

## THE COMPLEXITY OF THE STRINGS

The merging and separating of strings on the ultra subatomic scale (Planck scale) may be the controlling mechanism of the entire complex universe that we see around us. The functions of the most basic elementary particles can be explained as the vibrations of these strings. The complex diagram above represents mathematically the conglomerate interactions of strings.

Here, the process by which two particles exchange a force carrier (like a photon) can be envisioned much more simply by strings merging, exchanging energy, and moving apart again.

## MULTI-DIMENSIONED!

String theory suggests that there may be a number of unseen spatial dimensions in our universe that are all around us, but at the same time are so small that they are completely undetectable to us. Here is a simple diagram of a multi-faceted universe. If we were to shrink the image down, the spheres would appear as points. We would not be able to visually distinguish the spheres and we would have no knowledge of the complex dimensions hidden within them. Likewise, from a far off distance of a few hundred feet, a stretched out garden hose looks like a flat one-dimensional line. As we move closer to the hose, however, we quickly realize that it is a three-dimensional object. Superstring multi-dimensional theory works the same way – the other dimensions exist, but we can't see them from where we are.

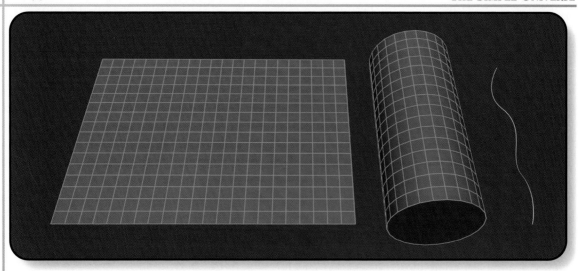

## THE HIDDEN DIMENSIONS

Superstring theory tells us that our universe may include several "rolled-up" tiny dimensions that are hidden within the four dimensions that we see around us. Above is a simple example of how multiple dimensions can exist within a single dimension: If you traveled east or west on the flat plane on the left, you'd eventually drop off the end of the world. But when that same plane is rolled into a cylinder, as on the right, you can travel east or west forever and never drop off – from that perspective, the end of the world doesn't appear to exist. Current superstring theories require a universe with nine or more dimensions. This multidimensional view has ruffled more than a few feathers in physics since we have yet to detect any of these other dimensions or even detect strings.

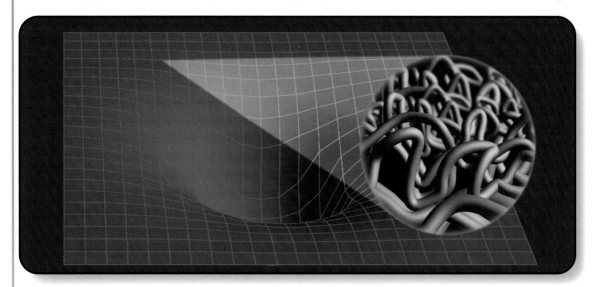

## BIZARRE TOPOLOGY

Where Einstein viewed the universe as a flexible membrane, pitted by gravity wells, modern quantum and string theories see the universe as a bizarre topology of shifting and ever-changing quantum surfaces. Our universe can be seen as a kind of quantum ocean. As an analogy, from high enough in the sky, the ocean appears to have a smooth surface, but the closer we get, the more clearly we see that the ocean is, in fact, a seething, foaming, undulating liquid.

## SO WHERE ARE THE STRINGS?

Although string theory does help to answer many questions that physicists have about the universe, it is still far from complete. The original string theories were unfortunately somewhat contentious, simply because there wasn't complete consensus among physicists at the more detailed levels. The result was the existence of no less than five competing versions of string theory. These theories have been labeled Type I, Type II, Type IIB, Heterotic O and Heterotic E. They are all fundamentally the same, however, with slight differences in the way that they deal with what physicists call supersymmetry. Supersymmetry is a principle that relates to specific properties of matter. The combination of supersymmetry and string theory leads to superstring theory.

In the mid-1990s, physicists came up with a new string theory, called M-theory. Each of the original five string theory's was seen as merely an aspect of a larger composite theory, much the same way that each point on a five-sided star is part of a larger unit. M-Theory states that the universe, as we understand it, should contain 11 dimensions (10 space and 1 time). M-Theory also predicts that, in addition to vibrating strings, there must be vibrating two-dimensional objects, known as membranes, and bizarre three-dimensional objects called branes. M-theory goes even further, predicting a number of even stranger undiscovered objects and particles in nature.

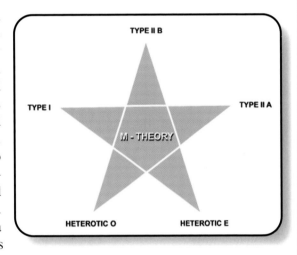

Although the M-theory version of superstring theory gives us a very compelling view of what the TOE may eventually look like, there is a problem – science does not currently possess the technology to experimentally prove this theory. M-theory makes the startling prediction that unification of the four forces (even gravity) should occur at 100,000,000,000,000,000,000 billion electron volts. That kind of energy cannot be manufactured by any machinery anywhere on the planet. This kind of energy could only have been found in the fires of creation. As it stands, modern science cannot confirm M-theory, or superstring theory in general. The final verdict is far from in on string theory. However, what is clear is that string theory has added yet another new twist in our quest to understand the universe. The high end mathematics produced by string theory continues to yield new truths about the universe. String theory may yet point the way to the Theory of Everything.

# CHAPTER 15: MULTIPLE UNIVERSES

Modern physics has buried within it the startling theory that our universe may not be the only one. There may be countless universes, each with its own properties. Although it may seem hard to believe, and even harder to conceive, our universe may be one of countless universes which may stretch over an infinite number of dimensions. Cosmology, combined with relativity, indicates that travel through distortions in space and time, caused by black holes, could lead to alternative universes running parallel to ours. Our theories, of our universe, may not apply to these parallel universes.

## JOURNEY THROUGH THE MULTIVERSE

The Big Bang may be responsible for the creation of not only our universe, but also an untold number of other universes. These universes may have expanded and evolved quite differently from our own and, as a result, may operate with bizarre physical parameters. The theoretical parallel universes may be so different from ours that, even if we found a way to travel into them, we may find them so alien that they would be totally inhospitable to any life from our universe.

## WORMHOLE SUBWAYS

Einstein's theories of relativity predict that, under certain special circumstances, a black hole may actually poke a hole into another part of our universe, or even into another universe altogether. Many scientists believe that wormholes are nothing more than science fiction, but there is a steady stream of growing data and analytical information that seems to substantiate the possibility of wormholes. Perhaps there will come a time when wormholes are used as subways, allowing matter, energy and the occasional traveler access to other universes.

## BLACK HOLE WOMB?

It has long been assumed that matter and energy falling into a black hole are doomed to be crushed down to an infinite density within the singularity. However, wormhole theory suggests that energy and matter perhaps might actually travel through a rip in the fabric of space-time and enter into another universe. Modern wormhole theory also allows for the startling possibility that matter and energy falling into a black hole, rather than falling into another universe, might instead create an entirely new universe. If, indeed, other dimensions do exist outside of those in our own universe (perhaps an infinite number of them), some or all of those additional dimensions are perhaps candidates for hosting new universes, waiting only for an initiating event to bring a new universe into existence. Matter and energy passing through a black hole from our universe into these other "available" dimensions may just provide the initiating event that starts a brand new universe rolling. So, black holes might be the wombs from which new universes are born. This theory further links the study of black holes to the birth of our universe. In fact, our own universe may have been created through this very process.

## ALIEN VISTA

A parallel universe may be governed by bizarre-seeming laws of physics, very different from those of our own universe. In a parallel universe, faster than light travel might be possible, and slower than light travel might be totally impossible. Or there may be a universe where gravity works in the opposite way to what it does in our universe – massive objects may actually repel each other. If we should find our Theory of Everything, it may well be uniquely ours – not applicable in any other universe where different physical laws apply.

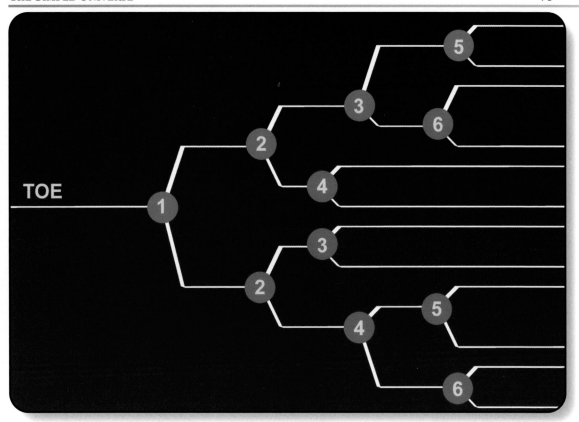

## TOE CODE BREAKER?

If there are any other parallel universes, an ultimate unified field equation, or Theory of Everything, might only explain our own universe, it may have no meaning in other universes whatsoever. Or it may act as a sort of code breaker, a mechanism allowing us to understand the basic principles that govern even radically different universes. The TOE may act in the same way that DNA does here on Earth – animals have radically different shapes and sizes, and exist in diverse environments; however, the fact is that the characteristics of all animals are based on the DNA within their genes. All DNA molecules employ the same double helix structure, and are centered around the same four chemical bases. In a similar fashion, the TOE might act as a blueprint for not just our universe, but others as well, no matter how radically different they may appear.

# CHAPTER 16: ON THE DARK SIDE

Cosmologists in the second half of the 20th century realized that they could calculate the mass of a galaxy by simply measuring the speed at which stars and gas clouds orbit about its center of rotation. In the same way, the mass of a cluster of galaxies can be determined by measuring the speed at which its component galaxies orbit the center of the cluster. However, when the masses of all of the visible galaxies are calculated, and added together, the total comes to roughly 10 times the mass of all of the stars, gas dust and debris visible in the entire universe. What cosmologists have concluded, after countless experiments, is that the majority of the matter and energy in our universe is not possessed by what we can see (and have photographed) in the sky. In fact, most of the mass of the universe is made up of invisible matter that science has aptly named "dark matter." So far, no one knows just what that dark matter is. However, there are some interesting theories – everything from garden variety black holes and neutron stars to exotic types of matter yet to be discovered.

## MACHOs

When evidence for dark matter was first detected, back in the early 1980s, it was assumed that it could be explained by some of the known but more bizarre astronomical objects like white dwarf stars, brown dwarfs, neutron stars, pulsars and black holes. These objects were dubbed MACHOs – Massive Compact Halo Objects.

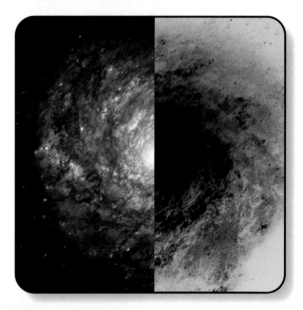

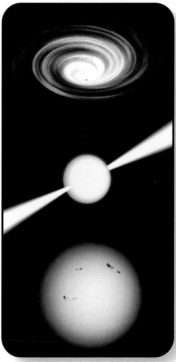

## WIMPs

More recent studies of galactic formation have led cosmologists to conclude that the major portion of the dark matter cannot, in fact, be in the form of MACHOs. Some physicists believe that this dark matter may be in the form of strange subatomic particles, like neutrinos. A growing number of scientists believe that this missing dark matter may consist of what physicists call WIMPs – Weakly Interacting Massive Particles. WIMPs are particles, as yet undetected, predicted by superstring theory and modern physics.

## EXPANSION, STABILIZATION OR COLLAPSE?

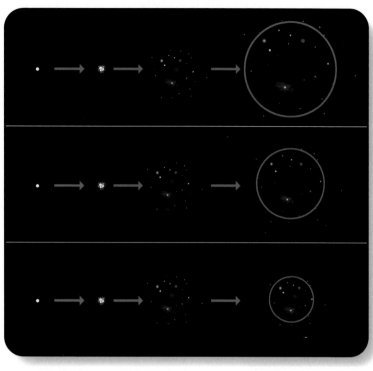

Cosmologists studying the universe believe that understanding what the dark matter is might help provide answers as to the ultimate fate of the universe. The theory goes something like this – Einstein and Hubble concluded, through mathematics and observation, that the universe has been expanding from a singularity for at least the last 15 billion years. If this is true, then the mere fact that the universe is expanding begs the following questions: will the universe expand forever, eventually thinning out and approaching absolute zero?; or will the universal expansion slow down, due to gravity, and stabilize at some equilibrium point?; or will gravity eventually win out over the expansion caused by the Big Bang and set the universe contracting, to eventually collapse back in on itself? These are all possibilities. One way to determine the fate of the universe requires us to add up all of the mass and energy in the universe. If the mass exceeds a certain calculable amount, gravity will eventually win out and start the universe contracting again. If the total mass is less than that, the expansion will continue. The possibility that the universe will expand to a certain size and then remain there is mathematically too small to be seriously considered.

## DARK MATTER AND STRING THEORY

Dark matter may be made up of yet undetected particles called WIMPs (Weakly Interacting Massive Particles). Although their exact properties are as yet unknown, their presence has been long predicted by superstring theory. Dark matter, MACHOs, WIMPs and superstring theory are all linked very closely together. Dark matter's apparent association with string theory ties it with the TOE. While an understanding of the TOE will tell us how the universe was formed, a comprehensive understanding of dark matter may tell us how it will end.

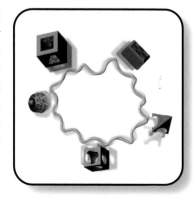

# CHAPTER 17: GOD AND THE THEORY OF EVERYTHING

The quest begun by early man and firmly envisioned by the Greeks continues to this day. However, there are those who believe that searching for an ultimate theory of the universe is futile, because it is an attempt to understand the mind of God. This kind of speculation comes from those who are deeply religious and from those who, for whatever reason, have an aversion to scientific explanations and mechanical views of the universe. However, a deeper scientific accounting of the genesis story may well lead us to a better understanding of the spiritual world. Finding a Theory of Everything may be important not only to better understand the workings of the universe, but also to understanding God and our place in creation.

## CAN THERE BE A CREATION WITHOUT A CREATOR?

This is a question which, if answered (if there is an answer), would redefine humanity for all time. Did the universe come into being through the actions of a creator or God? Or is our universe here by random chance, having evolved out of pure nothingness? Until recent times, all humanity saw our universe, and all life within the universe, as the creation of some divine being, who created us, and all that we see around us, with some higher purpose in mind. But there is mounting theoretical evidence that perhaps our universe is the product of random chance.

# HUMAN EXPERIENCE VS. UNIVERSAL TRUTH

Human experience teaches us that if there is a creation there must be a creator. We all have fathers and mothers who created us, and they in turn had fathers and mothers. The chair you sit on did not spontaneously come into being; people had to design it and build it. This book that you hold in your hands didn't just magically appear; it was the result of actions by people and machines. Human experience teaches us that every creation must have had a creator. But modern physics teaches us that what we see can be a deception. Objects are not solid; solidity is an illusion caused by the interaction of subatomic forces. Our perception that the universe was created by a creator might also be an illusion – based on human experience in the natural world, and not on scientific fact.

# RATING THE GAME!

Those in the scientific community with strong religious roots often argue that even if proof existed that a universe can appear out of nothingness, that does not automatically discount the existence of God. God may play a dual role in the origin of the universe. God might be both the observer and the observed. In other words, what if the creator, or God, is not only the one responsible for the creation of the universe, but in fact, is the universe. In a more practical sense, Einstein saw matter and energy as interchangeable; the same applies to space and time; perhaps God is similarly transposable with the universe, and the two cannot be separated, but are different aspects of the same thing.

# CHAPTER 18: FROM ONE SIMPLE VIEW TO ANOTHER

Our current scientific view of the universe, based on quantum theory and general relativity, represents the most complete understanding humanity has ever had of the world we see around us. Throughout recorded history our view of the universe has been guided – and tainted – by human biases. Many early humans, who were closer to the Earth than their descendants, saw the creation and evolution of the universe simply as biological facts; they felt no burning need to know whence they came. Later humans ascribed the creation of the universe to human-like gods and goddesses, who created the world solely that their creations might worship them. Even today's scientific views may be unduly affected by our desire to find a simple underlying theory to explain the workings of the universe. Many scientists who don't share this simplified universe theory wonder "why should our universe be explainable using simple and rational explanations?" Indeed, those who study quantum theory often point to the overly complex underpinnings of the quantum realm. Many see the quantum world as anything but simple, and kin to chaos. It may be that our scientific views of the universe are no more valid than the ancient views, now considered quaint by modern man.

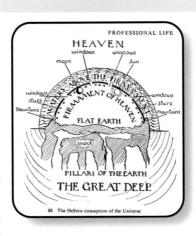

88 The Hebrew conception of the Universe

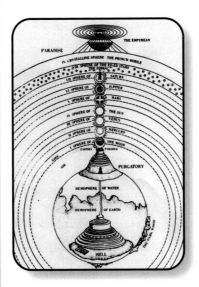

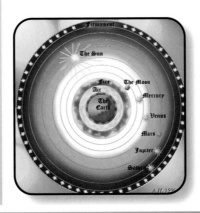

## SEEING THE TRUTH!

Although we may chuckle at the views of the ancients, marveling at the many ways in which they viewed the universe, we must remember that our ancestors were doing exactly the same thing as today's scientists and philosophers – trying to explain the universe and how it works. And in most respects, our view of the universe today is not that much different from that of those who came before us. And like earlier humans, we construct our models of the universe based on observation and experimentation. When there are gaps in our knowledge that observation and experimentation cannot fill, we make logical guesses, and uphold those guesses until such time as we can prove or disprove them. Most earlier societies did likewise – made their educated guesses, and stuck by them until better information became available. The only differences in how we proceed today revolve around our more sophisticated tools, and a greater wealth of previously accumulated knowledge – accumulated by the same people whose earlier theories we have supplanted. Our desire to understand the secrets of the universe is at least as old as our recorded history, and probably much older.

Tycho Brahe and other astronomers and scientists of the renaissance period helped to construct our modern view of the universe. Brahe, Galileo, Newton and Halley were renegades – and pioneers in their fields. They risked the wrath of the authorities of their day in order to bring new enlightened ways of thinking to the rest of the world. It's interesting to note that many of the modern heroes of physics and astronomy were once considered charlatans by the scientific communities of their times.

## GOING BACK TO THE BEGINNING!

Purveyors of many of the world's religions have often throughout history been at odds with the scientific leaders of their times over the nature and origin of the universe. In many ways, religious and scientific accounts of our origin are quite similar, not totally contradictory at all. In the Judeo Christian beliefs, the universe is without form until the word of God creates light and starts the process rolling, an "event zero" from which all else proceeds. A similar theme, with alternative names and places, seems common to many religions – Creation involves a simple act, executed by an infinitely complex entity, which spawns the entire universe. Oddly enough, science explains the origin of the universe in a very similar manner: the breaking of perfect symmetry, at the Big Bang, starts a series of cascading events that lead to our universe. In both scenarios, ultimate power acts upon maximum simplicity to bring into being a universe of continually increasing complexity.

## PARALLEL UNIVERSES?

Reviewing our current scientific picture of the universe we find: string theory, suggesting that there may be several more (yet to be detected) spatial dimensions within our own universe; quantum theory; relativity; and astronomy. And they all allow for the existence of other, parallel universes. A simple visualization of this would be our universe as a giant bubble, floating in a sea of foam. Each bubble within the foam is a separate, self-contained universe, each with its own unique properties. Each

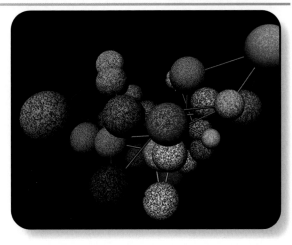

universe, like each bubble, is separated from all of the others, and can expand and evolve completely independently of the others. The possibility exists that, like a big splash in the foam, the Big Bang that created our universe may have given birth to other universes as well.

## CREATION THROUGH DIVISION?

The study of black holes and the search for grand unified field equations (the Theory of Everything) are closely coupled. The universe is thought to have begun in a state of simplicity comparable to that found at the heart of a black hole – a massive singularity and a unique event. But where did the singularity and the event come from? Scientists seriously consider it possible that matter and energy passing out of the universe by way of a black hole can provide the initial singularity and Big Bang-type event that create a new universe. In simplified form, the thinking is that: matter and energy captured by a black hole cannot escape it; their cumulative effects serve to further increase the gravitational force of the black hole (since its mass is increasing); at some point, the black hole's effective mass increases to the point were the gravitation is great enough to cause a further collapse; this new collapse causes a portion of the black hole's accumulated mass to actually fall in on

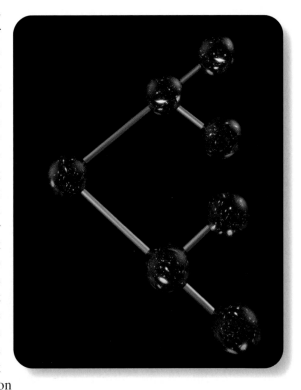

itself – it literally becomes "pinched-off" and separates from our universe, becoming a new universe, in contact only with itself; with the environment of its black hole "parent" no longer confining it, the "daughter" universe begins expand and cool.

## UNIQUE BALANCE

Each universe in the picture has its own unique properties. Our universe may be destined to die through a fiery Big Crunch, where all the matter in the universe falls back in on itself in a mirror image of the Big Bang. Or, the universe might possibly end in a Big Freeze, with everything continuing to expand until the densities of energy and matter are virtually zero. Some have suggested that, if other universes exist, we may be able to escape to them and survive the end of our universe. However, the properties of our universe may be so radically different from those of other universes that we may not be able to survive in them.

For example, the universe we escape into may have no such thing as a strong force, so that we would immediately dissolve into a soup of quarks. We might also enter into a universe made up of antimatter and instantly be annihilated. Traveling to other universes may be a more dangerous prospect than at first thought.

## CHAPTER 19: THE HIGGS FIELD

Modern physics uses experimentation, theory and high end mathematics to try to explain the universe around us. One of the more unique and interesting predictions in recent physics is the proposal of the Higgs field. The Higgs field is believed to be responsible for creating the mass possessed by particles. Some physicists call the Higgs field "the missing force," and theorize that it may be carried by an exchange particle, as is the case with electro-magnetism, the strong force and the weak force.

### THE HIGGS FIELD

According to one theory in modern physics, our universe may be immersed in a giant energy field called the Higgs field. Interaction with this field is what gives particles their mass. Different particle types interact with the Higgs field with different strengths, and therefore possess differing masses. (Some particle types, such as photons, have no mass – they don't interact with the Higgs field.)

The Higgs force is theoretically carried by a proposed exchange particle called the Higgs boson. The Higgs boson gets its mass, like all other particles, by interacting with ("swimming in") the Higgs field. But as you can imagine, the Higgs boson differs from all of the other particles that we've seen. It can be thought of as a dense spot in the Higgs field (like a drop of water in a cloud of water vapour) which can travel around, like any other particle. Higgs bosons "cling" to other particles that they encounter; the more Higgs bosons that cling to a particle, the heavier the particle is.

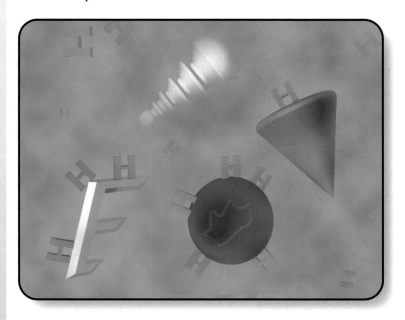

# HIGGS FIELD AND THE HIGGS BOSON

The Higgs field and the Higgs boson can be viewed functionally as the same thing. We can use the sky as a metaphor for the Higgs field. Although we may not be able to see them, our atmosphere is filled with tiny water vapor droplets; we only see these droplets when they attach themselves to dust particles and accumulate to form clouds. The Higgs field is all around us; however, it only becomes visible when portions of it execute its Higgs boson function. Much like the four forces, the presence of the Higgs field can only be recognized through the actions it performs. In this sense, we can say that the Higgs field and the Higgs boson are the same thing.

# HIGGS FIELDS AND PARTIES

The Higgs field gives mass to elementary particles. The more the Higgs field drags on a given particle, the larger that particle's mass becomes, and the greater is the amount of energy required to move it. As an analogy, if you are at a party and Brittany Spears enters the room, there is a good chance that she will attract many of the guests towards her. Getting from one side of the room to the other may prove difficult for her. However, you may easily move through the room, because you attract fewer people to you (you are less massive).

# THE STANDARD MODEL AND THE HIGGS FIELD

Although we have yet to detect a Higgs boson, or confirm the existence of the Higgs field, the Standard Model of physics does predict its existence. Without the Higgs field, there can be no mass to any particle in the universe. There would be no stars, planets, moons, or anything else for that matter. Many physicists believe that with the more powerful accelerators being constructed at Fermi Lab and CERN it is only a matter of time before evidence is found for the existence of the Higgs Field. Its existence will bring science much closer to the unification prophesied by the Theory of Everything.

# CHAPTER 20: PUTTING TOGETHER THE LINES OF INQUIRY

The search for a Theory of Everything has led scientists down many paths. It is becoming ever more apparent that many different lines of inquiry may have to be traveled before we can arrive at the ultimate truth that explains the universe. By following and combining these multiple lines of inquiry we may acquire a better overall understanding of the universe. If we can understand how these various lines of inquiry unite, we may finally be able to answer the most complex questions about our universe.

## ANSWERING THE BIG QUESTIONS

Science doesn't simply focus on one avenue at a time looking for answers, but examines several fields concurrently. In the search for a TOE, scientists have ventured down many roads, some by accident, some by design. Each of these scientific roads illustrates, in its own way, aspects of the physical laws that control the universe we live in. As we better understand these individual components of the universe, and how they interact, we gain a better understanding of the overall framework that holds the universe together.

## CONFLICT AMONG THE LINES OF INQUIRY

One of the problems with our current theories of the universe is that results from different lines of inquiry are sometimes in apparent conflict with one another. As mentioned earlier in the book, one of our major goals is to combine Einstein's general relativity, which describes interactions in the macro universe, with quantum theory, which concerns the behavior of subatomic particles. Physicists are hoping to some day soon be able to unite our understanding of gravity with quantum theory to create a new discipline: quantum gravity. The unification of gravity and quantum theory will take us a great deal closer to a Theory of Everything.

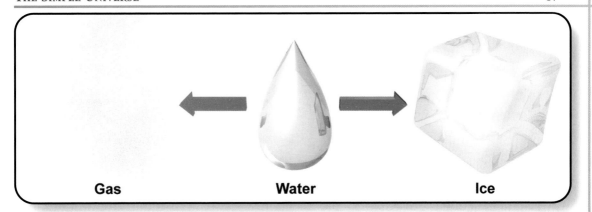

| Gas | Water | Ice |

## THE RELATION OF MATTER AND ENERGY

The correlation between matter and energy is well understood in modern physics. However, there are many gaps in our knowledge. Further understanding of strange phenomenon like dark matter, strange matter and the Higgs field may force us to adjust and perhaps even discard some of our old notions of the material world and contemporary physics.

## MULTI-DIMENSIONAL PHYSICS AND THE THEORY OF EVERYTHING

What we see in the physical world is often an illusion – our interpretation of what our eyes tell us is limited by the scale of our eyesight. At greater levels of detail, much of what we call solid appears very different. A smooth sheet of paper seen through a microscope appears as a mountainous landscape. If we could reduce our perspective to the atomic scale, we would see that what our 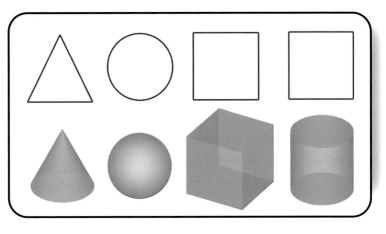 eyes had previously labeled as solid was, in fact, anything but. Our everyday, four-dimensional world might just be an illusion as well. String theory, among other branches of physics, suggests that our universe might actually include at least nine dimensions, perhaps more! In some manner that we don't yet understand, the matter and energy around us might also exist in, and have effects upon, multiple additional, unseen dimensions within our universe.

## FUTURE PARTICLE THEORY

The combining of various diverse lines of inquiry in modern physics has lead to some startling predictions. For example, the merger of supersymmetry and string theory (to form superstring theory) has introduced the possibility that our universe may contain many more unknown types of exotic particles. What cosmologists call "dark matter" might be the manifestation of these previously unknown particles. As we continue to link together previously independent lines of investigation, we will no doubt be further amazed at how exotic but beautiful our universe truly is.

# CHAPTER 21: THE CONTINUING SEARCH

So, after all of this, where do we stand in the grand scheme of things? Thanks to Albert Einstein, and a handful of other people who came before him, we are on a continuing journey to find ultimate simplicity – a theory to explain, in rational mathematical terms, every event that has occurred in our universe since its beginning. It is a quest to find perfect symmetry, a symmetry that has not existed since the Big Bang.

The current state of the art in modern physics tells us that there are four fundamental forces in nature; under normal conditions, these four forces act independently of one other and each has its own unique properties. However, if we could increase the equilibrium thermal energy (turn up the universe's ambient temperature) to the point where conditions resembled those found in the early universe (some 15 billion years ago) we would find the properties of the four forces changing with increasing energy. As the ambient energy passes a certain threshold, the changing properties of the electromagnetic and weak forces will have come to be the same, and they can be considered as a single – electroweak – force. At another, higher threshold, the electroweak and strong forces will have acquired matching properties and can be treated as a single force – the electro-nuclear force. Finally, scientists feel confident that at a final, yet higher threshold, the electro-nuclear force and gravitation will have acquired the same properties, resulting in a single superforce, whose properties would be defined by a set of unified field equations – our sought-after Theory of Everything.

Science has developed some remarkable theories attempting to explain how some of these forces merge, but has yet to come up with a blanket theory explaining the unification of all four forces. There is promise, however, in what has been called modern string theory or M-theory. It appears that M-theory, and string theory in general, have within them the concepts that could serve to unite all of nature's forces together into a single common framework, helping to explain the workings of the universe, its origin, and perhaps even its ultimate fate. At the time of this writing, physicists are looking hard at M-theory, hoping that an ultimate Theory of Everything will emerge from its elucidation, to answer our many questions about the universe.

If we are intelligent enough to discover our Theory of Everything, we'll find that it's not just an equation, or a scientific theory, but is, at root, an utterly simple and rational idea – a concept pertaining to everything in our universe. It will be such a simple idea that we will know immediately upon its discovery that this is what Einstein was searching for. And it will be so compelling that we will ask ourselves, "how could it have been otherwise?"

## COMMON GOALS

There's a classical conception of scientists held by laymen – lonely people working long hours into the night, trying to solve impossible equations, and propounding impossibly complex theories. Einstein, Newton, Galileo and many other scientists of the past did, indeed, work like this. They were unable to find like-minded people to share in their journeys of discovery. Today, however, scientists work not as individuals, but in large groups – vast networks of creative people, pooling their resources pursuing common goals. Places like CERN and Fermi lab employ thousands of people and rely on support networks and universities spread across the entire planet. When Carlo Rubbia won the Noble Prize in physics for his confirmation of the electroweak force, he accepted the award not just for himself, but on behalf of the countless people who had supported his effort. This is science at its best; no longer are scientists confined by outdated ideologies, religious boundaries, color, age, sex or nationality. Today's science includes the products of countless people from every background.

## UNBREAKING BROKEN SYMMETRY

If our modern view of the universe is correct, then our universe owes its existence to a breaking of perfect symmetry. For example, during the electroweak era of the universe ($1/10$ billionth of a second after the Big Bang), the weak bosons and the photon had similar properties and both acted as the exchange particle for the electroweak force. Immediately following this period, the electroweak force separated into the nuclear weak force and the electromagnetic force. The electromagnetic exchange particle is the massless photon, whereas the exchange particle for the weak force is a very massive weak boson. The mass difference between these two types of exchange particles is typical of the broken symmetry that occurred as the early universe expanded. In today's universe, physical laws show a consistent preference for conservation and balance, yet the mass difference between weak bosons and photons exists.

The universe is composed of matter that condensed out of the ambient energy released during the Big Bang, much the same way that snow flakes appear to form out of thin air or ice freezes out of a running stream in winter – matter congealing out of energy. Each atom in the universe is a little storehouse of energy – a battery charged by the Big Bang – continually emitting and absorbing energy as it interacts, through photons, with other atoms. In the same way, light reaching us from every star, and heat from the sun, are energy moving from place to place within our universe – energy released into the universe by the Big Bang and still in the universe today.

## SEARCHING FOR WHAT WE KNOW!

Is there a single truth, a single physical law that dominates the universe? Many scientists believe so and have devoted their lives to finding that universal law. Albert Einstein, one of the modern pioneers in the search, once said, "the most incomprehensible thing about the universe is that it is comprehensible."

When we look, both at the subatomic level, and out to the stars, we see evidence of the simplicity that binds our entire universe together. We see evidence everywhere that the universe grew out of a single potent seed of energy, smaller than the smallest atom. And in a sense, we were all there – every bit of matter that makes us up, every atom in our bones, blood, and brains, every scrap of matter on Earth can trace its lineage back to the origin of the universe. The physical laws that govern our universe came into being in the fires of its genesis, and have held sway ever since. We are all composed solely of matter and energy that have been a part of our universe since its birth; this makes us all brothers and sisters in a very real sense. We are all partners in our vast, wonderful universe. The story of the universe is far from over; it is evolving all around us, its ultimate destiny is still unknown. But with the dedicated, cooperative efforts of so many intelligent men and women working with faithful determination towards unlocking the secrets of our universe, it is surely only a matter of time before we realize our Theory of Everything.

What physicists call broken symmetry was absolutely essential for the universe, as we know it, to exist. Theory suggests that the universe began in a state of perfect symmetry, which ceased to exist almost immediately, as things expanded and cooled. The properties of matter and energy, as they exist today, are a consequence of this broken symmetry. The physical laws that made this flower possible were written in the first unimaginably small fraction of a second following the Big Bang.

# CONCLUSION

Science has come a long way, since the thoughts of the early Greeks, in an attempt to understand the long road from the beginning of the universe until the present. There has been a tendency for the leading scientific figures of each age to believe that all of the important answers had been found. Lord Kelvin (1824-1907), perhaps the most esteemed scientific leader of his time, is known to have said, "There is nothing new to be discovered in physics now; all that remains is more and more precise measurement." Subsequent events have always shown otherwise.

Today we are of a different mind. We know fully that we are nowhere near the end of discovery; there will always be more to learn, new events to understand. Our quest to combine General Relativity and quantum theory is still ongoing. Our sought-after "Theory of Everything" will not, of course, cover literally "everything." This Unified Field Theory, as it is also called, will bring us a long way towards more fully understanding the universe, but there will always be more. Consider gravity: everything we know today, even including Einstein's marvelous General Relativity, can only describe the characteristics of gravity's behavior. But what *is* gravity? How and why does it function? We are as far from understanding the essence of gravity as Newton was.

So, what else is on the horizon? What other wonderful concepts are awaiting us on the continuing journey to find the Theory of Everything? 100 years ago the idea of quarks would not have been given serious consideration. 50 years ago the idea of string theory would not have been entertained. What about 50 and 100 years into the future? What new concepts will be generally accepted then that have not even been imagined today. It may well be that as we continue to acquire understanding, at an ever-increasing rate, we may discover that there are increasingly more marvels in the universe yet to be understood.

Considering all that we have learned from it, the Theory of Everything can be considered a "means" as much as an "end." And as we've seen, the basic concepts that we have shared in this book are simple enough for all of us to understand. At the same time, however, we can see that the "simple universe" is anything but.

# APPENDIX 1 –
# PARTICLES IN THE STANDARD MODEL OF PHYSICS

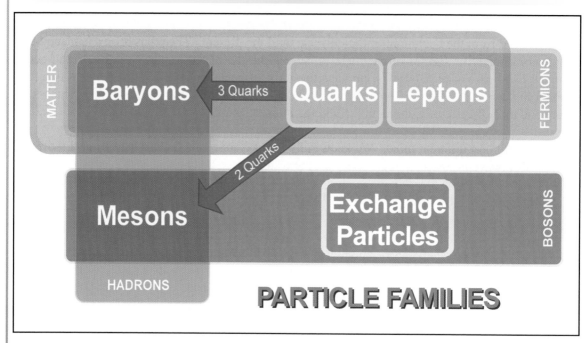

In the Standard Model of physics, subatomic particles are divided into two groups – fermions and bosons. Fermions are the particles that make up matter, while bosons are the exchange particles (force carriers) of the fundamental forces. The Standard Model encompasses electromagnetism, the nuclear weak force and the nuclear strong force. Gravity is not part of the Standard Model.

Note: The supersymmetry counterparts of the particles described herein are not discussed. They are not generally considered to be part of the Standard Model of Physics.

## FERMIONS

Fermions are the particles that compose the matter in the universe. There are two elementary particle types in the fermion group – leptons and quarks. Leptons are particles that stand alone, whereas quarks exist only in groups, with two or three quarks binding together to form larger particles called hadrons (see below).

All fermions have a spin value of ½ and are therefore constrained by the Pauli exclusion principle (see below). Spin describes the nature of the energy distribution in a collection of the particles.

### Leptons
There are six types (called flavors) of lepton, the most familiar of which is the electron (see Table of Elementary Particles). There is a corresponding antilepton for each of the six leptons.

### Quarks
There are six types (called flavors) of quark (see Table of Elementary Particles). Quarks do not exist in isolation; quarks and gluons are bound together in particles called hadrons (see below). There is a corresponding antiquark for each of the six quarks.

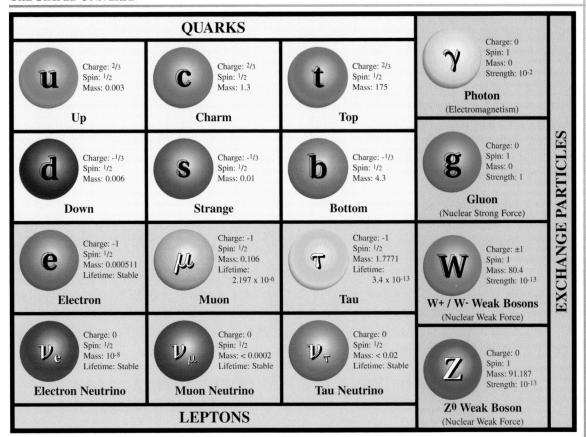

**Table of Elementary Particles**

## BOSONS

The bosons are the force exchange particles. The photon (electromagnetic force) and gluon (nuclear strong force) are massless elementary particles, while the weak bosons (nuclear weak force), also called intermediate vector bosons, are very massive elementary particles.

Mesons (not elementary particles) are also a type of boson; mesons are extremely short-lived hadrons (see below) that are the exchange particle for an aspect of the nuclear strong force called the "residual strong interaction," which is responsible for binding protons and neutrons together into atomic nuclei.

Some bosons also contribute mass to the matter with which they interact (the mass-equivalent of their binding energy also contributes mass).

All bosons have a spin value of 1 and are therefore not constrained by the Pauli exclusion principle (see below). Spin describes the nature of the energy distribution in a collection of the particles.

## HADRONS

Quarks and gluons are bound together into particles called hadrons. There are two types of hadrons – baryons and mesons. Hadrons are the particles that interact by the nuclear strong force, and most hadrons have extremely short lifetimes, existing only as intermediate byproducts of physical processes such as particle collisions.

## Baryons

Baryons are massive particles which are made up of three quarks, the most familiar of which are the proton and the neutron. All baryons are fermions (matter particles). There is a corresponding antibaryon for each baryon, except those which are electrically neutral (no charge), each of which is its own antiparticle. There are about 120 types of baryons, most of which have extremely short lifetimes.

## Mesons

Mesons are particles of intermediate mass which are made up of two quarks – a quark-antiquark pair. All mesons are bosons (exchange particles). There is a corresponding antimeson for each meson, except those which are electrically neutral (no charge), each of which is its own antiparticle. There are about 140 types of mesons, all with extremely short lifetimes.

## PAULI EXCLUSION PRINCIPLE

No two particles (of the same type) in an atom can have identical "quantum numbers." This principle applies to all fermions (but not bosons).

This simple principle is a decisive factor in the nature of our universe. The fact that no two electrons in an atom can have identical quantum numbers is directly responsible for the fact that there are unique elements (types of atoms) in our universe, as given in the periodic table of elements (see below).

## PROPERTIES OF FUNDAMENTAL FORCE INTERACTIONS

| | ELECTRO-MAGNETIC | WEAK NUCLEAR | STRONG NUCLEAR | | GRAVITY |
|---|---|---|---|---|---|
| | | | Fundamental | Residual | |
| **Acts on:** | Electric Charge | Flavor [1] | Color Charge [2] | [3] | Mass – Energy |
| **Particles affected:** | Electrically charged | Quarks, Leptons | Quarks, Gluons | Hadrons | All |
| **Strength:** | $7.3 \times 10^{-3}$ | $10^{-5}$ | 1 | | $6 \times 10^{-39}$ |
| **Range:** | Infinite (inverse square law) | $10^{-17}$ meters (diameter of a proton) | $10^{-15}$ meters (diameter of a medium size nucleus) | | Infinite (inverse square law) |
| **Exchange Particle:** | Photons | Weak bosons ($W^+$ $W^-$ $Z^0$) | Gluons | Mesons | Graviton (not yet observed) |

**Properties of the Fundamental Force Interactions**

[1] The weak interaction is the only process in which a quark can change to another quark, or a lepton to another lepton – the so-called "flavor changes."

[2] The force between quarks is called the color force. In some respects, it can be thought of as the "source" of the strong interaction..

[3] Inside a baryon, the color force has properties different from interactions between nucleons, and is responsible for the confinement of the quarks..

## PERIODIC TABLE OF ELEMENTS

Each element in nature is comprised of a unique type of atom, e.g., gold is made of gold atoms and oxygen gas is made of oxygen atoms. The term compound (as opposed to element) refers to any substance containing more than one type of atom, e.g., water $H_2O$ is a compound having both hydrogen and oxygen atoms. In nature, it is extremely rare for a pure element to exist – any quantity of the element will contain impurities, even if only in microscopic amounts.

Periodic Table of Elements legend:

- Non-Metals
- Alkali Metals
- Alkali Earth Metals
- Transition Metals
- Other Metals
- Halogens
- Noble Gases
- Rare Earth Metals

State key:
- N — Gas
- V — Solid
- Br — Liquid
- Tc — Synthetic

Main table (atomic number shown with symbol):

| 1 | 2 | 3 | 4 | 5 | 6 | 7 | 8 | 9 | 10 | 11 | 12 | 13 | 14 | 15 | 16 | 17 | 18 |
|---|---|---|---|---|---|---|---|---|---|---|---|---|---|---|---|---|---|
| H (1) Hydrogen | | | | | | | | | | | | | | | | | He (2) Helium |
| Li (3) Lithium | Be (4) Beryllium | | | | | | | | | | | B (5) Boron | C (6) Carbon | N (7) Nitrogen | O (8) Oxygen | F (9) Fluorine | Ne (10) Neon |
| Na (11) Sodium | Mg (12) Magnesium | | | | | | | | | | | Al (13) Aluminum | Si (14) Silicon | P (15) Phosphorus | S (16) Sulfur | Cl (17) Chlorine | Ar (18) Argon |
| K (19) Potassium | Ca (20) Calcium | Sc (21) Scandium | Ti (22) Titanium | V (23) Vanadium | Cr (24) Chromium | Mn (25) Manganese | Fe (26) Iron | Co (27) Cobalt | Ni (28) Nickel | Cu (29) Copper | Zn (30) Zinc | Ga (31) Gallium | Ge (32) Germanium | As (33) Arsenic | Se (34) Selenium | Br (35) Bromine | Kr (36) Krypton |
| Rb (37) Rubidium | Sr (38) Strontium | Y (39) Yttrium | Zr (40) Zirconium | Nb (41) Niobium | Mo (42) Molybdenum | Tc (43) Technetium | Ru (44) Ruthenium | Rh (45) Rhodium | Pd (46) Palladium | Ag (47) Silver | Cd (48) Cadmium | In (49) Indium | Sn (50) Tin | Sb (51) Antimony | Te (52) Tellurium | I (53) Iodine | Xe (54) Xenon |
| Cs (55) Cesium | Ba (56) Barium | La (57) Lanthanum | Hf (72) Hafnium | Ta (73) Tantalum | W (74) Tungsten | Re (75) Rhenium | Os (76) Osmium | Ir (77) Iridium | Pt (78) Platinum | Au (79) Gold | Hg (80) Mercury | Tl (81) Thallium | Pb (82) Lead | Bi (83) Bismuth | Po (84) Polonium | At (85) Astatine | Rn (86) Radon |
| Fr (87) Francium | Ra (88) Radium | Ac (89) Actinium | Rf (104) Rutherfordium | Db (105) Dubnium | Sg (106) Seaborgium | Bh (107) Bohrium | Hs (108) Hassium | Mt (109) Meitnerium | Uun (110) Ununnilium | Uuu (111) Unununium | Uub (112) Ununbium | Uuq (114) Ununquadium | | Uuh (116) Ununhexium | | Uuo (118) Ununoctium | |

Lanthanides and Actinides:

| 58 | 59 | 60 | 61 | 62 | 63 | 64 | 65 | 66 | 67 | 68 | 69 | 70 | 71 |
|---|---|---|---|---|---|---|---|---|---|---|---|---|---|
| Ce (58) Cerium | Pr (59) Praseodymium | Nd (60) Neodymium | Pm (61) Promethium | Sm (62) Samarium | Eu (63) Europium | Gd (64) Gadolinium | Tb (65) Terbium | Dy (66) Dysprosium | Ho (67) Holmium | Er (68) Erbium | Tm (69) Thulium | Yb (70) Ytterbium | Lu (71) Lutetium |
| Th (90) Thorium | Pa (91) Protactinium | U (92) Uranium | Np (93) Neptunium | Pu (94) Plutonium | Am (95) Americium | Cm (96) Curium | Bk (97) Berkelium | Cf (98) Californium | Es (99) Einsteinium | Fm (100) Fermium | Md (101) Mendelevium | No (102) Nobelium | Lr (103) Lawrencium |

**Periodic Table of Elements**

The "atomic number" of each element (shown in each upper right corner) is a whole number equal to the number of protons in its nucleus, which is also equal to the number of electrons the atom possesses. This balance in the number of protons and electrons makes the atom as a whole electrically neutral. Whenever an atom loses or gains one or more electrons, it is referred to as an "ion" rather than an atom. Ions have a net electrical charge.

## Plasma

Plasma, often called the fourth state of matter, is a volume of gas comprised of charged ions and neutral atoms, in which the positive ions and negative ions within the volume are equal, so that the overall charge of the volume is neutral. We do not generally encounter plasma in our everyday lives, but the fact is that in the universe as a whole, more than 99% of all matter is believed to be plasma.

## Isotopes

The number of neutrons in an atom is approximately (or exactly) equal to the number of protons. Unlike protons, however, the number of neutrons can vary. Looking at iron as an example – all iron atoms have 26 protons and most, but not all, iron atoms have 30 neutrons. Specifically, naturally occurring iron atoms exist in the following proportions:

| Number of Neutrons | Percent of Total |
|---|---|
| 28 | 5.845 |
| 30 | 91.754 |
| 31 | 2.119 |
| 32 | 0.282 |

So, most iron atoms have 26 protons plus 30 neutrons in their nuclei. The different configurations (neutron counts) of the atom are called "isotopes." As shown in the table, there are four naturally occurring isotopes of iron. The atomic weight (mass) cited for an element is a composite value taking all of its isotopes and their relative proportions into account.

# APPENDIX 2 – THE COMPONENTS OF MATTER

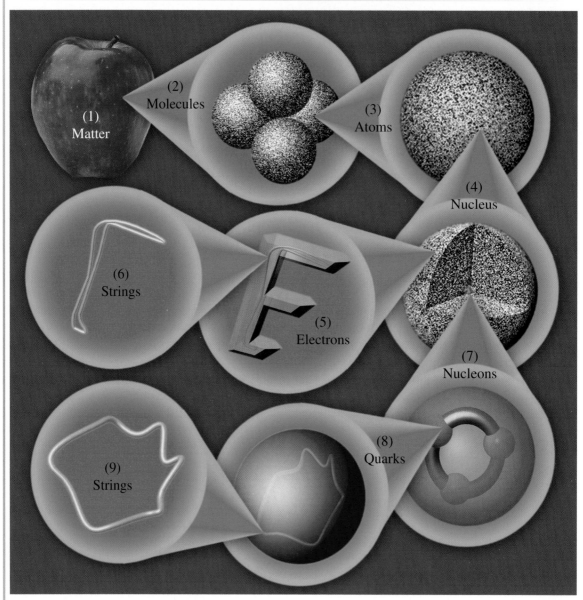

Science has investigated continuously smaller scales trying to define the basic constituents of matter. The diagram above illustrates the components of matter, according to modern physics, as summarized below.

(1) **Matter** – The everyday objects that we see have macro properties, as perceived by our five senses, that determine how we interact with them.

(2) **Molecules** – Everything that we encounter in our world is structured as molecules. The well-known water molecule, $H_2O$, includes two hydrogen atoms and one oxygen atom, making water a compound (more than one type of atom). Oxygen gas, $O_2$, on the other hand, as it appears in our atmosphere is an element (contains only one type of atom).

(3) **Atoms** – Since the time of the ancient Greeks, the atom was believed to be the smallest component of matter. Only in the last 200 years has science gone beyond that paradigm to propose and then detect subatomic particles.

(4) **Nucleus** – Up until the last half century, the atom was seen as composed of a nucleus plus electrons orbiting the nucleus. As science advanced, protons and then neutrons were added to the model, making the nucleus a composite item rather than a simple particle.

(5) **Electrons** – The electron has always been considered as an elementary particle. We know now that an electron can absorb or emit energy in the form of photons, but the electron itself is still basically an indivisible elementary particle. The electron is a member of the particle family called leptons.

(6) **Strings** – Contemporary string theory suggests that the elementary particles are perhaps not particles at all, but one-dimensional "strings." String theory is, as yet, purely a mathematical model, with no experimental evidence to substantiate it, but it does explain some aspects of particle physics that the particle model can not.

(7) **Nucleons** – The nucleus of the atom is made up of nucleons – protons and neutrons. Although much more massive than electrons, the nucleons were also, up until the last half century, considered to be indivisible elementary particles. We know now that a nucleon is, in fact, a combination of three lesser particles called quarks. The nucleons are a type of particle called baryons, which are part of the hadron group.

(8) **Quarks** – Quarks are elementary particles that combine to make larger particles called baryons (3 quarks) or mesons (2 quarks). Quarks are indivisible particles and are bonded together by gluons, the nuclear strong force exchange particles. Quarks cannot exist in isolation.

(9) **Strings** – As with the electron, it has been proposed that quarks are actually one-dimensional strings, rather than actual particles.

# APPENDIX 3 – THE THEORY OF EVERYTHING

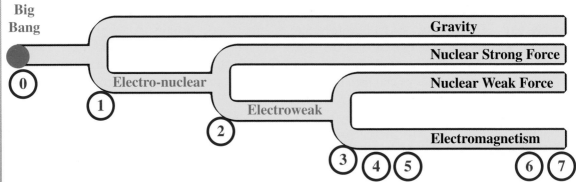

Albert Einstein was the first to propose what he called a "unified field theory," a set of physical laws that would merge all four of the fundamental forces of nature into a single "superforce," which could be described by a single, simple set of rules. Contemporary physicists believe that a merger of the quantum forces (nuclear strong, nuclear weak, and electromagnetic forces) with general relativity (which accounts for the workings of gravity) would yield the desired unified field theory, or as some call it, a Theory of Everything.

There have been many attempts to merge the quantum forces with gravity, but none have yet succeeded. However, physicists believe that M-theory (the most recent version of string theory) has within it the concepts that will lead to the unification of gravity and the quantum forces. The introduction of M-theory has perhaps put us on the very door step of the Theory of Everything.

The following table (referenced to the diagram above) is a summary of the major events in the universe following the Big Bang. All values presented are, of course, theoretical, but events from (3) onward have experimental evidence to support them.

| | Time (seconds) | EVENT | Temperature (°K) | Particle Energy |
|---|---|---|---|---|
| (0) | 0 | **Big Bang** | | |
| (1) | $10^{-44}$ | The Unified Field force uncouples to the Electro-nuclear and Gravitational forces. | $10^{32}$ | $10^{19}$ GeV |
| (2) | $10^{-37}$ | The Electro-nuclear force uncouples to the Electroweak and Nuclear Strong forces. | $10^{28}$ | $10^{15}$ GeV |
| (3) | $10^{-10}$ | The Electroweak force uncouples to the Nuclear Weak and Electromagnetic forces. | $10^{15}$ | 100 GeV |
| (4) | $10^{-5}$ | Quark confinement begins. Quarks can no longer exist in isolation, but are bound together (by gluons) into hadrons. | $10^{12}$ | $10^{-1}$ GeV |
| (5) | $10^{2}$ | Neutrino transparency occurs. The fireball from the Big Bang has cooled enough to become transparent to neutrinos. | $10^{9}$ | $10^{-4}$ GeV |
| (6) | 3 x $10^{5}$ years | The cosmic background radiation, left over from the Big Bang, becomes transparent. | 3 x $10^{3}$ | 3 x $10^{-10}$ GeV |
| (7) | 12 x $10^{9}$ years | Present time. | 2.7 | 2.3 x $10^{-13}$ GeV |

# GLOSSARY

**Accelerator**
See "Particle accelerator."

**Alpha Centauri**
Generally referred to as the closest star to Earth, Alpha Centauri is actually a triple star system some 4.3 light-years from Earth. One member of the triplet, Proxima Centauri, is actually the closest at 4.22 light-years. (See also Proxima Centauri.)

**Andromeda Galaxy**
A mid-sized galaxy whose closest stars are some 2.3 million light-years away from us. The Andromeda galaxy is estimated to contain more than 300 billion stars. Andromeda is the closest major galaxy to our own galaxy, the Milky Way.

**Antimatter**
Antimatter has properties similar to normal matter, except that all electrical charges are reversed. For example, a proton is positively charged, but an antimatter proton (an antiproton) is negatively charged. When matter and antimatter come into contact, the result is mutual annihilation.

**Astronomy**
Astronomy deals with the origin, evolution, composition, distance, and motion of all extraterrestrial objects. Astrophysics, which studies the physical properties of extraterrestrial objects, is a component of Astronomy.

**Astrophysics**
The study of the physical properties (physics and chemistry) of extraterrestrial objects.

**Atom**
Atoms are considered to be the basic building blocks of all matter. An atom has a nucleus (protons and neutrons) surrounded by clouds of orbiting electrons. Each elemental substance has its own unique atom, defined by the number of protons in its nucleus (oxygen atoms have 8 protons, calcium atoms have 20 protons, and gold atoms have 79 protons). All of the known atoms are presented in the Periodic Table of Elements.

**Baryons**
Baryons are particles comprised of three quarks. Examples of baryons are protons and neutrons. Baryons are fermionic hadrons.

**Big Bang**
The initiating event generally believed to have brought our universe into being. The Big Bang was the explosion of an infinitely dense singularity, the remnants of which then cooled and expanded into the universe we see today. The Big Bang is believed to have occurred somewhere between 15 and 20 billion years ago. Also known as Event Zero.

**Big Crunch**
A theory stating that the universe may stop expanding and then begin collapsing back in on itself, reforming an ultimate singularity in the reverse process of the Big Bang.

**Big Freeze**
A theory stating that the universe will keep on expanding until all matter and energy have decayed to zero density. This is the condition of maximum entropy.

## Black Hole

A collapsed star with such a strong gravitational attraction that nothing, not even light, can escape from it. A black hole appears black because light cannot be reflected from it.

## Boson

One of two classes of elementary particles, bosons include the exchange particles (force carriers) – photons (electromagnetic force), $W^+$, $W^-$ and $Z^0$ weak bosons (nuclear weak force) and gluons (nuclear strong force). The boson class also includes mesons, which are the exchange particle for the "residual strong force," the component of the nuclear strong force responsible for binding protons and neutrons into the nucleus of the atom. (See also Fermions.)

## Carlo Rubbia

The Italian physicist who first conceived of using matter / antimatter collisions to simulate conditions in the early universe. The application of his theory, in massive particle accelerators like CERN, led directly to the confirmation of the theorized electroweak force in 1983.

## CERN

A particle accelerator and research facility that sits on the border between France and Switzerland. CERN gained international recognition in 1983 when a team of physicists and mathematicians lead by Carlo Rubbia of Italy verified the existence of the electroweak force (electromagnetic and nuclear weak forces unified). Carlo Rubbia and his CERN team won the Noble Prize in physics for their work.

## Cosmic Microwave Background Radiation

Microwave radiation permeating the entire universe. This is a residual energy of the Big Bang, which has been cooling and thinning over time. It is the ambient radiation level of the universe.

## Cosmology

Cosmology is the scientific study of the large scale properties of the universe as a whole. It uses astronomy, astrophysics, particle physics and mathematics to explore the origin, evolution and ultimate fate of the entire Universe. There are several proposed "cosmologies" within the field. Their basic differences revolve around the existence and inclusion of other universes, and whether our universe is open or closed. In its very broadest sense, the "cosmos" includes our universe and all other universes (past, present and future) whether they are theoretically accessible to/from ours or not.

## Dimension

A direction of movement through space or time. Physics, as we understand it, presently tells us that our universe is made up of four known dimensions, three of space (length, width and height) and one of time. Modern superstring theories suggest that there may be as many as 10 or 11 dimensions.

## Edwin P. Hubble

An American who discovered, in 1929, that the universe is expanding. With his telescopic observations of the night sky, Hubble helped to confirm Einstein's theory that the universe was indeed expanding. NASA paid Hubble tribute in 1990 by naming their new orbiting space telescope platform after the astronomer who pioneered deep space astronomy.

## Electromagnetic

Dealing with or referring to the electromagnetism.

## Electromagnetism

One of the four fundamental forces of nature. Electromagnetism includes magnetism and all forms of vibrational energy from radio waves and microwaves, through light and sound, right up to x-rays and gamma rays. The exchange particle (force carrier) for Electromagnetism is the photon.

## Electron

Electrons are the subatomic particles comprising the outer shell of all atoms. Electrons have a negative charge, very little mass, and belong to the class of elementary particles called leptons. Electrical current results from the transfer of electrons, from atom to atom, through a conductor between two points of differing charge.

## Electro-nuclear Force

A theoretical superforce that would result from the unifying of the electroweak force and the nuclear strong force, as described by the grand unified theory (GUT). The theorized electro-nuclear exchange particle (force carrier) is the X particle.

## Electroweak Force

A unified force existing when the electromagnetic force and the nuclear weak force acquire similar properties in a high ambient energy (high temperature) universe, such as existed very briefly following the Big Bang. The electroweak force was theorized several years prior to its confirmation in 1983 by scientists working with the CERN particle accelerator.

## Energy

Energy is a measure of the capability of an object or system to do work when the system undergoes change. Work is defined as exerting a force through a distance.

## Enrico Fermi

An Italian physicist who predicted, in 1933, that there was a nuclear weak force acting within the confines of the atom. Fermi Lab in the United States is named for him.

## Epoch of Light

An era some one million years after the Big Bang marked by the emergence of light. At this point, the universe has thinned out enough that photons (the carriers of the electromagnetic force) can travel through space without running into particles (protons, neutrons etc.). This is also the time when the first atoms were formed.

## Event Zero

An alternative name for the Big Bang. (See also Big Bang.)

## Fermi Lab

A large particle accelerator and science lab outside of Chicago on the Illinois plains. Fermi Lab is named for Enrico Fermi, who first proposed the nuclear weak force interacting within the core of the atom. Fermi Lab has a string of successive finds that have helped to revolutionize our view of the universe. Fermi Lab detected the existence of the bottom quark (1977), top quark (1995) and the tau neutrino (2000).

## Fermion

One of two classes of elementary particles. Fermion particles make up nearly all of the visible matter in the universe. Fermions include leptons, quarks, and baryons. Unlike bosons (the other class), fermions have a uniqueness in that no two fermions within an atom can have exactly the same properties; they must differ in the value of at least one property (the Pauli Exclusion Principle). The four forces in nature are the result of fermions exchanging bosons.

## Four Forces

There are four fundamental forces that govern the behavior of all matter and energy in the universe. These four forces are: electromagnetism, the nuclear weak force, the nuclear strong force, and the gravitational force. Each of these forces is carried by its exchange particles (or bosons as they are more properly known). The exchange particles are photons (electromagnetic force), weak bosons (weak force), gluons (strong force) and gravitons (gravity). Contemporary physicists assert that the properties of each of these forces will change as the ambient energy level (temperature) increases. At sufficiently high temperatures, the properties of different forces will become the same and the forces can be considered to have merged into a single new force. This force integration is outlined in the unified field theories. The Standard Model of physics does not at present include the graviton, since it is still a completely theoretical particle.

## Galaxy

A system of millions or billions of stars maintained as a group, separate from other galaxies, by their mutual gravity. Many galaxies also contain large gas clouds (which give birth to stars and planets). The vast majority of galaxies are billions of years old and may have super-sized black holes at their cores. Our galaxy, the Milky Way, is a mid-sized galaxy and contains approximately 250 billion stars and is about 100,000 light-years across.

## Galileo Galilei

A 17th century Italian astronomer and logician who is considered by many to be the father of modern science. Galileo pioneered the use of the telescope for astronomical observations. His theories about the nature of the universe aroused the wrath of the Catholic Church, which placed him under house arrest and forbid him from teaching his theories. Galileo died while under house arrest. His theories about the universe and his lines of inquiry were later used by Sir Isaac Newton and others to establish much of our modern scientific understanding of the universe.

## General Theory of Relativity

A theory developed by Albert Einstein in 1915 that describes how matter and energy behave in the presence of a gravitational field. Einstein's earlier Special Theory of Relativity (a limited case scenario which ignored gravity) predicted that time does not flow at a fixed rate – the faster you move, the slower time appears to flow (this effect only becomes significant at velocities close to the speed of light). The General Theory of Relativity includes the effects of gravitation, and predicts that gravity (the curvature of space-time by matter) will stretch or shrink distances (depending on their direction with respect to the gravitational field) and also will appear to slow down or "dilate" the flow of time.

## GeVs (Giga Electron Volts)

An electron volt (eV) is the kinetic energy gained by an electron when it is accelerated through a one volt electric potential. One GeV is $10^9$ (or 1,000,000,000) electron volts. These units are used by physicists to describe the energy levels of subatomic particles.

## Gluon

The exchange particle (or boson) that carries the nuclear strong force.

## Grand Unified Theory (GUT)

Any of several theories attempting to unite electromagnetism, the nuclear weak force and the nuclear strong force into one superforce. This united force is known as the electro-nuclear force.

## Gravitation

One of the four fundamentals forces of nature. Gravitation is the force responsible for the mutual attraction that a mass has for all other masses. Quantum physics suggests that gravitation is carried by an exchange particle (or boson) called a graviton. As yet, no one has experimentally detected the existence of the graviton. All things being equal, gravity is actually the weakest of the four forces, but its effects are proportional to the masses involved, and planets have a lot of mass.

## Graviton

The exchange particle (or boson) which carries the gravitational force. Although theorized by quantum physics and string theory, gravitons have not yet been detected.

## Hadrons

Hadrons are a class of particles in which quarks and gluons are confined and cannot be separated. Hadrons include baryons and mesons. (See also Baryons and Mesons.)

## Higgs Boson

According to one recent theory, our universe may be immersed in a giant energy field called the Higgs field. Interaction with this field is what gives particles their mass. Different particles interact with the Higgs field with different strengths, and the mass of any given particle is proportional to the strength of its interaction. The hypothesized Higgs field is carried by an exchange particle called the Higgs boson.

## Hubble Space Telescope

An orbiting space telescope platform placed in a 381 mile orbit in the spring of 1990. Because the Hubble telescope is above Earth's atmosphere, it can take undistorted, farther reaching pictures of the universe. The space telescope is named after legendary astronomer Edwin P. Hubble who determined, through observation, that the universe was expanding.

## Inflation

A theory stating that the very early universe underwent a period of enormous rapid expansion that overpowered gravity's inclination to have the universe collapse back in on itself. Theoretically, during the inflation period of the universe only gravity and the electro-nuclear force existed.

## Sir Isaac Newton

Often referred to as the grandfather of modern physics; Isaac Newton is most famous for his work chronicled in *The Principia*. In his book, *The Principia*, Newton explained in detail, for the first time, the laws of gravity. First put forth in 1687, the principles outlined in *The Principia* are still in use today and, although superseded for precise determinations by General Relativity, Newton's laws of gravity are still invaluable in understanding the universe.

## Leptons

A class of subatomic particles (from the fermion family) that are not affected by the nuclear strong force. The leptons include electrons, three types of neutrino, muon and tau particles. There are also six anti-lepton types, one for each lepton.

## Light-year

The distance traveled through a vacuum by light in one year, equivalent to about six trillion miles. Light travels (in a vacuum) at 186,000 miles per second, usually denoted as $c$. Light travels only slightly more slowly through our atmosphere (or any other transparent non-vacuum).

## Local Group

A group of local galaxies within 10 million light-years of the Earth. The local group consists of our home galaxy, the Milky Way, its next door neighbor, the Andromeda galaxy, and a handful of lesser galaxies close to our own.

## M-Theory

A more recent interpretation of string theory that combines the original five string theories into a single framework. M-theory is still being explored and is not yet completely understood. However, it appears that M-theory might provide the foundation for a Theory of Everything.

## Mass

A measure of the total amount of material in an object. Mathematically, mass is determined by the forces acting on the object (such as gravity) and its acceleration state.

## Mesons

Mesons are the exchange particles for the residual strong force, the component of the nuclear strong force that binds protons and neutrons into the nucleus of the atom. Mesons are particles created from the combination of two quarks, one of which is always an antimatter quark. Mesons are extremely short lived.

## Multiverse

Most factions in modern cosmology allow for the existence of universes other than our own. The Multiverse is the name given to the entire collection of universes. There is not currently consensus among cosmologists as to whether any or all of these other universes, should they exist, are accessible to/from our own. The term "parallel universes" is often used to denote these other universes.

## Neutrino

A particle with no charge, and little or no mass that moves at the speed of light. Neutrinos have the ability to move through great volumes of very dense material (like a planet) without affecting or being affected by what they pass through. Neutrinos are in the lepton class of subatomic particles.

## Neutron

Neutrons, with protons, make up the nucleus of the atom. They are very massive subatomic particles and have no charge. Neutrons are made up of three quarks (one "up" and two "down" quarks).

## Neutron Star

Neutron stars are small collapsed stars comprised almost exclusively of neutrons. Unlike their pulsar cousins, neutron stars do not spin rapidly.

## Nuclear Strong Force

One of the four fundamental forces of nature. The strong force is responsible for binding quarks together to form particles, such as protons and neutrons, and is also the binding force that keeps protons and neutrons together in the nuclei of atoms (overcoming the electromagnetic mutual repulsion of the protons). The exchange particle (force carrier boson) for the strong force is the gluon. Mesons act as the exchange particle for the residual strong force.

## Nuclear Weak Force

One of the four fundamental forces of nature. The weak force is responsible for the decay of particles into smaller particles (changes the identity of particles). The weak force acts upon all known fermions. It governs the creation and interaction of the elementary particles known as neutrinos, among others. One consequence of this particle decay is nuclear fusion (as takes place in stars); another is certain types of radioactivity.

## Nucleus

The central core of the atom. The nucleus is the home of the protons and neutrons (the nuclei of all atoms have both protons and neutrons, except the hydrogen nucleus, which is a single proton). The nuclear strong force acts to bind the protons and neutrons into a nucleus against the electromagnetic repulsion of the charged protons.

## Particle Accelerator

A machine used to accelerate subatomic particles to near light speed and then collide them either with a stationary target or other particles (or antimatter particles). Particle accelerators range from tiny hand-sized units to instruments that are miles in circumference.

## Photon

An exchange particle (also called a force carrying particle or a boson) which carries the electromagnetic force. Photons travel at light speed.

## Physics

The scientific study of the interaction of matter and energy.

## Positron

The antimatter counterpart of the electron; it has a positive charge (as opposed to the electron's negative charge) but otherwise has the same properties as an electron.

## Proton

Protons, with neutrons, make up the nucleus of the atom. They are very massive subatomic particles and have a positive charge. Protons are made up of three quarks (two "up" quarks and one "down" quark).

## Proxima Centauri

The star Proxima Centauri (of the Alpha Centauri triple star system) is the closest (extrasolar) star to Earth; it is 4.22 light-years away. This means that light takes 4.22 years to get from Proxima Centauri to Earth, and that we are seeing it as it was more than four years ago – we are effectively looking more than four years back in time.

## Pulsar

A pulsar is a rapidly spinning neutron star that ejects intense radiation along its magnetic axis. Twice each rotation, as the magnetic axis points towards our section of the sky, this radiation is very briefly visible from Earth, providing the pulsing effect. (See also neutron star.)

## Quanta

According to quantum physics, the smallest packet of energy that an electron can give off or absorb when changing energy levels (you can't have an energy packet of 1½ quanta, only whole numbers). This dictates, therefore, that any type of energy transfer (such as the action of photons) can only involve whole numbers of quanta.

## Quantum Physics

The branch of science dealing with the realm of the subatomic. Quantum physics deals with the smallest components of matter and energy, their relation to each other, and their effects on the macro universe that we see.

## Quark

Quarks are fundamental matter particles which combine in groups of two or three to form mesons or baryons, respectively. Mesons and baryons make up the group called hadrons, the class of particles in which quarks and gluons are confined and cannot be separated. There are six different types (called flavors) of quarks – up, down, charm, strange, top and bottom.

## Quasar

Quasars are thought to be energetic galaxies with massive black holes at their centers, going through a fairly violent youthful phase. Quasars are found in the farthest regions of the universe (as far as 15 billion light-years) and most quasars seem to have formed during the early days of the universe.

## Radiation

Energy propagated in the form of subatomic particles (photons) or electromagnetic waves. It includes all forms of vibrational energy. Electromagnetism is the force driving all radiation, from radio waves through to gamma rays.

## Relativity

See General Relativity and Special Relativity.

## Singularity

A point of extremely high or (essentially) infinite density that occupies zero space. The most common example is the center of a black hole, but small (submicroscopic) singularities can theoretically exist temporarily in free space.

## Special Relativity

Albert Einstein's special theory of relativity predicted that time does not flow at a fixed rate – the faster you move, the slower time appears to flow (this effect only becomes significant at velocities close to the speed of light). A direct result of special relativity is the fact that nothing traveling below the speed of light can be accelerated to the speed of light. The second important aspect introduced by special relativity is the realization that there are no absolute quantities in the universe, every measurement and every perception is relative only to a specific observer. While being one of the most important developments ever in the history of physics, special relativity fell short of being an all-encompassing theory because it did not take gravity into account. (See also General Relativity.)

## Standard Model (of Physics)

This is the name given to the collected theories of particle physics (quantum theory). It asserts that all matter is comprised of combinations of 12 types of subatomic particles (6 quarks and 6 leptons) and includes the interactions of three of the four fundamental forces (electromagnetic, nuclear weak and nuclear strong). Gravity and the graviton exchange particle are not considered as part of the Standard Model. Six types (flavors) of quark (plus their corresponding anti-quarks) are the building blocks for heavier particles – mesons, made of two quarks, and baryons (including protons and neutrons), made of three quarks. Electrons are one of six types of particle called leptons, which, like quarks, are indivisible fundamental particles. In the Standard Model, quarks and leptons interact through the exchange of photons, gluons, and the W and Z bosons.

## String Theory

String theory postulates that the smallest indivisible elements of nature are tiny, one-dimensional vibrating strands called strings. Strings have yet to be detected and have only been deduced through high end mathematics and theoretical physics. Conventional physics teaches us that all matter in the universe is made up of atoms. These atoms in turn are made up of very small elementary particles. String theory takes things a step further, asserting these elementary particles are themselves made up of ultra, ultra small strings, which exist at the tiniest levels possible. The vibrations of these strings, according to theory, are the root events of every event in the universe.

## Strong Force

See Nuclear Strong Force.

## Superstring Theory (SST)

String theory conveniently explains the properties of bosons (particles composed of two quarks). To also encompass the fermions (heavyweight baryons, composed of three quarks; and lightweight leptons, such as electrons, that are not composed of quarks), physicists have combined supersymmetry with string theory to introduce superstring theory. (See also Supersymmetry.)

## Supernova

A rare event in which a super-massive star, at the end of a its lifetime when its nuclear fuel is exhausted, blows itself apart in a spectacular explosion. The blast wave ejects the star's envelope far out into space. Supernovas are so powerful that many will briefly outshine the galaxies they dwell in.

## Supersymmetry

Supersymmetry theory contends that each exchange particle has a massive "shadow partner" particle with fermion-like properties. Likewise, quarks and leptons would each have a boson-like "shadow partner" particle. The net result would be that the effective mass of each pair – the combination of a particle plus its shadow – would serve to correct the observed mass imbalances that occur when considering the particles without their "shadows" (e.g., weak bosons are very massive whereas photons are effectively without mass).

## Theory of Everything (TOE)

Also called unified field theory, this sought-after theory, and its corresponding equation(s), that will comprehensively encompass all of the physical laws of the universe. The TOE will be the ultimate unified field theory, describing all four of the fundamental forces and all types of matter, and their interactions. The TOE will knit general relativity and quantum theory into one unified field theory. Experimentation has supported currently accepted theories in physics which have the nuclear weak force and the electromagnetic force combined into the electroweak force in a high energy (high temperature) environment. It is generally accepted that at even higher energy, the electroweak and nuclear strong forces will similarly unite into what has been called the electro-nuclear force. The inclusion of gravity at yet higher energy has been proposed, but is still considered somewhat speculative.

## Unified Field Theory

See Theory of Everything (TOE).

## Virgo Super Cluster

A vast archipelago of galaxies which stretches over tens of millions of light-years. The Local Group of galaxies is but a tiny section of the Virgo super cluster. The Virgo Super Cluster itself is one of thousands of super clusters that spread across the known universe.

## Weak Boson

The weak bosons, also called intermediate vector bosons, are the exchange particles (force carriers) for the nuclear weak force. There are three types of weak boson – $W^+$, $W^-$ and $Z^0$, and there are also subtypes of these three. Photons, gluons and gravitons are other examples of bosons. (See also Bosons.)

## Weak Force

See Nuclear Weak Force.

## X Particle

A theoretical particle which is believed to exist at extremely high energy levels, such as were present in our universe immediately following the Big Bang. The X particle would be the exchange particle (force carrier) of the electro-nuclear force, the theoretical unification of the electroweak force (unified electromagnetism and nuclear weak force) and the nuclear strong force.

# ABOUT THE AUTHOR

Ian Brewster is a science enthusiast and writer. He has been actively studying physics and astronomy from a very young age. He specializes in cosmology and physics and has written several papers on the subject for science journals and magazines. Ian is also a full time freelance writer and has written for a variety of magazines on every subject from women's issues to geopolitics, military Special Forces, world history, computers and the martial arts. An avid martial artist, Ian has spent over half of his life in the study of the combat arts and currently owns and operates his own training gym in Toronto, Ontario, Canada. Ian is available for seminars and is always glad to answer questions from the general public. He can be reached at the following email address: i_brewster@yahoo.com.

# ABOUT THE GRAPHIC ARTIST

Ken Shiwram is and has always been dedicated to the arts in many different forms, from drama to painting. He has made creating art his lifetime passion and has earned various awards in the art field. Ken believes that a true passion from within will never die, unless it is neglected enough to be wiped out of existence. Besides art, Ken has another passion, nature. He is also a freelance illustrator and art instructor. He can be reached via email at: kshiwram@hotmail.com.

# SUGGESTED READING

*A Brief History of Time*, by S. Hawking, Bantam, 1988.
*The Elegant Universe*, by Brian Greene, Vintage Books, 2000.
*Hyperspace*, by Michio Kaku, Anchor Books, 1994.
*The Emperor's New Mind*, by Roger Penrose, Oxford University Press, 1989.